The 

t

GCSE

Mathematics Revision

Higher Tier

CAMBRIDGE
UNIVERSITY PRESS

PUBLISHED BY THE PRESS SYNDICATE OF THE UNIVERSITY OF CAMBRIDGE
The Pitt Building, Trumpington Street, Cambridge CB2 1RP, United Kingdom

CAMBRIDGE UNIVERSITY PRESS
The Edinburgh Building, Cambridge CB2 2RU, United Kingdom
40 West 20th Street, New York, NY 10011–4211, USA
10 Stamford Road, Oakleigh, Melbourne 3166, Australia

First published 1997

Printed in the United Kingdom at the University Press, Cambridge

A catalogue record for this book is available from the British Library

ISBN 0 521 57907 4

This book has been written and compiled by

Eric Gower
Ray Harris
Spencer Instone
Elizabeth Jackson
Paul Scruton

The authors' warm thanks go to the teachers who tested the draft materials in their schools and to Howard Baxter and William Wynne Willson who gave advice from an examiner's standpoint.

Certain questions in this book are reproduced by kind permission of the following:

The Midland Examining Group
The Northern Ireland Council for the Curriculum Examinations and Assessment
The Southern Examining Group
London Examinations: a division of Edexcel Foundation
(formerly University of London Examinations and Assessment Council)
The Welsh Joint Education Committee

These questions are acknowledged individually in the text. None of the above groups bears any responsibility for the accuracy or method of working in example answers to these questions.

Contents

Shape, space and measures

Formula sheet

Area of triangle = $\frac{1}{2} \times \text{base} \times \text{height}$

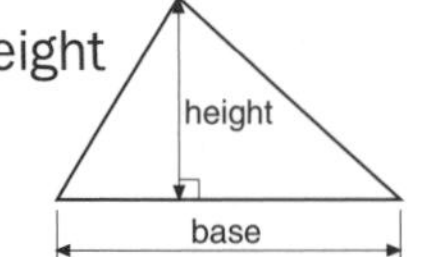

Area of parallelogram = base × height

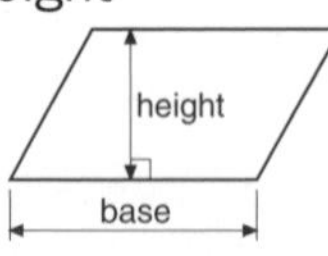

Area of trapezium = $\frac{1}{2}(a+b)h$

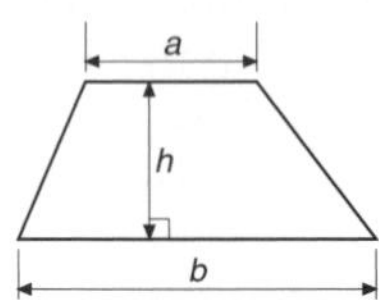

Volume of cuboid = length × width × height

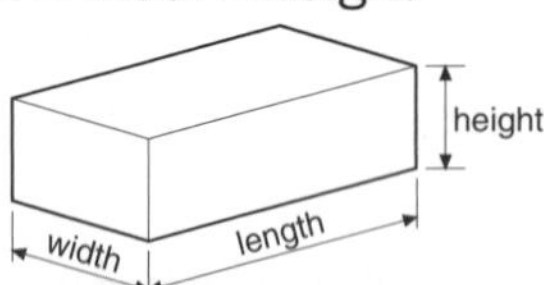

Volume of prism = (area of cross-section) × length

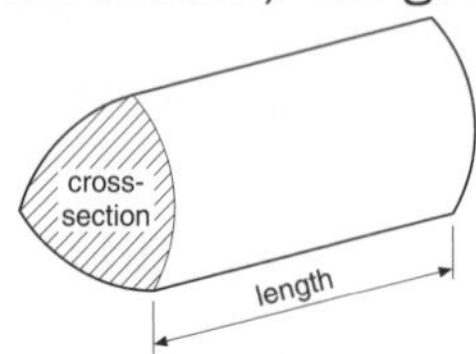

Volume of cylinder = $\pi r^2 h$

Curved surface of cylinder = $2\pi rh$

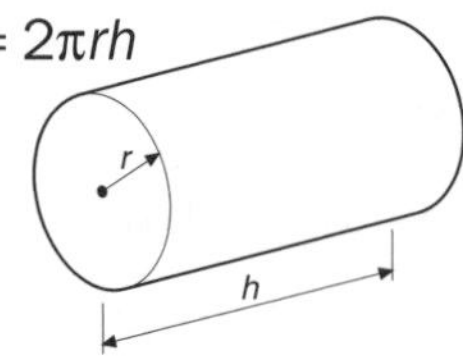

Volume of sphere = $\frac{4}{3}\pi r^3$

Surface area of sphere = $4\pi r^2$

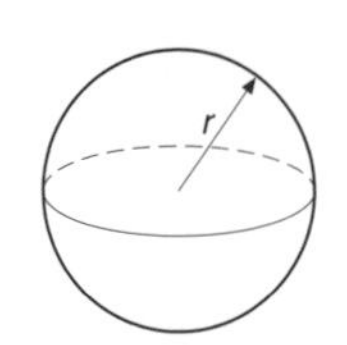

Volume of cone = $\frac{1}{3}\pi r^2 h$

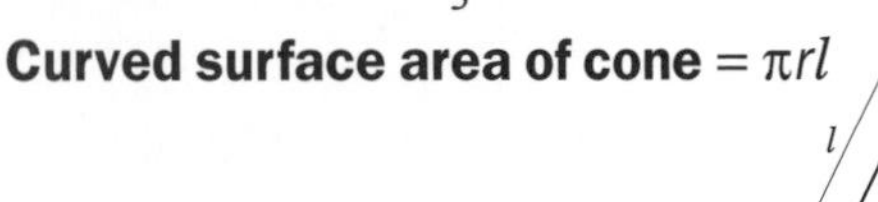

Curved surface area of cone = $\pi r l$

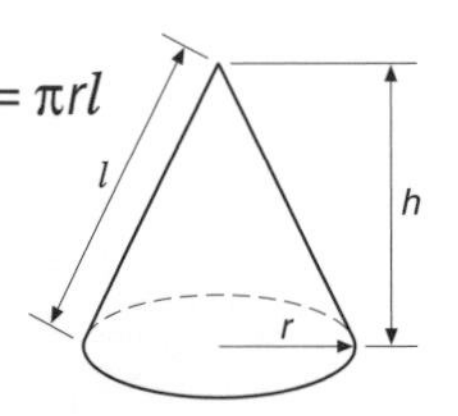

Pythagoras' Theorem

$a^2 + b^2 = c^2$

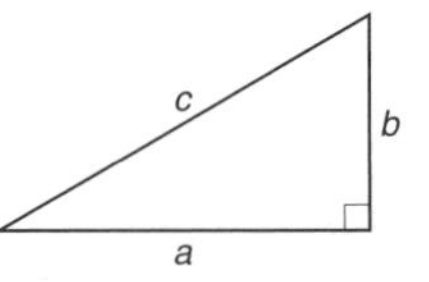

Circumference of circle = $\pi \times$ diameter

= $2 \times \pi \times$ radius

Area of circle = $\pi \times (\text{radius})^2$

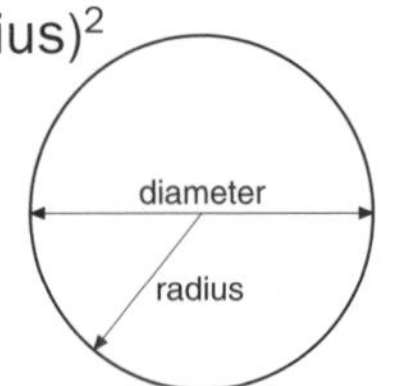

Trigonometry

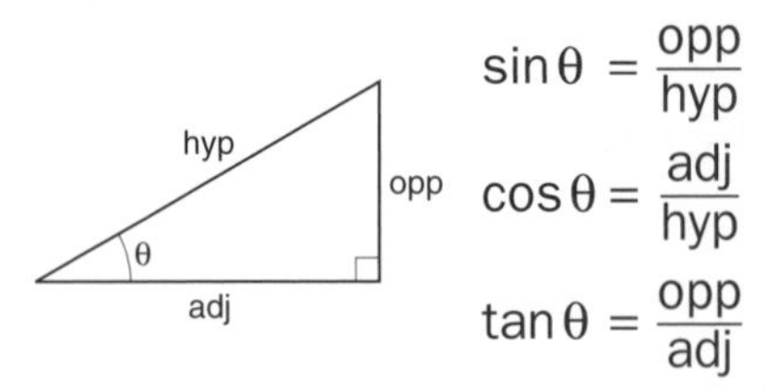

$$\sin\theta = \frac{\text{opp}}{\text{hyp}}$$

$$\cos\theta = \frac{\text{adj}}{\text{hyp}}$$

$$\tan\theta = \frac{\text{opp}}{\text{adj}}$$

In any triangle ABC

Sine rule $\quad \frac{a}{\sin A} = \frac{b}{\sin B} = \frac{c}{\sin C}$

Cosine rule $\quad a^2 = b^2 + c^2 - 2bc\cos A$

$$\cos A = \frac{b^2 + c^2 - a^2}{2bc}$$

Area of triangle = $\frac{1}{2}ab\sin C$

C, b, a, A, B, c

The quadratic equation

The solutions of $ax^2 + bx + c = 0$ where $a \neq 0$, are given by

$$x = \frac{-b \pm \sqrt{(b^2 - 4ac)}}{2a}$$

Standard deviation

Standard deviation for a set of numbers $x_1, x_2, \ldots, x_n$, having a mean of $\bar{x}$ is given by

$$s = \sqrt{\left(\frac{\Sigma(x-\bar{x})^2}{n}\right)} \text{ or } s = \sqrt{\left(\frac{\Sigma x^2}{n} - \left(\frac{\Sigma x}{n}\right)^2\right)}$$

How to use this book

This book covers the content of all the GCSE mathematics syllabuses at higher level. There is a brief explanation of each topic followed by questions (including many from past GCSE examinations) for you to work through. You will find answers and hints to all the questions at the back of the book.

How you use this book depends on how much you need to revise. It is divided into sections – Number, Algebra, Shape, space and measures and Handling data. You could start at the beginning of a section and work through it steadily or you could pick out the things that you are unsure about (with the help of the contents pages) and concentrate on them. If you need further help, references to SMP 11–16 books, at the end of each set of Answers and hints, tell you where to find it.

Some questions need worksheets. Ask your teacher for these.

NUMBER

Properties of numbers

The **factors** of a whole number are those numbers which divide exactly into it.
(The factors of 20 are 1, 2, 4, 5, 10 and 20.)

The numbers 3, 6, 9, 12 and 15 are **multiples** of 3; 3 is a factor of each of them.

A number which has only two factors, itself and one, is called a **prime number**.
(2, 3, 5 and 7 are all prime numbers, but 1 is not.)

Numbers can be split into the **product** of prime factors. ($20 = 2 \times 2 \times 5$ and $70 = 2 \times 5 \times 7$)

1, 4, 9, 16, 25, ... are **square numbers** and 1, 8, 27, 64, 125, ... are **cube numbers**.

Make sure you agree with these **square** and **cube roots**:

$\sqrt{36} = 6$ $\quad \sqrt[3]{125} = 5$ $\quad \sqrt{1{\cdot}44} = 1{\cdot}2$ $\quad \sqrt[3]{8000} = 20$

To find the **reciprocal** of a number you divide 1 by the number.

For example: the reciprocal of 6 is $1 \div 6 = \frac{1}{6}$.

the reciprocal of $\frac{1}{6} = 1 \div \frac{1}{6} = 1 \times \frac{6}{1} = 6$

the reciprocal of $\frac{2}{5} = 1 \div \frac{2}{5} = 1 \times \frac{5}{2} = \frac{5}{2}$

Make sure you can find squares, cubes, square roots and cube roots on your calculator, and can use the reciprocal key – usually [1/x] *or* [x^{-1}].

Dividing by a fraction ► page 3

1 (a) Write down all the factors of 28.
(b) Write 28 as the product of prime factors.
(c) Write down all the prime numbers between 40 and 70.
(d) Write down three numbers which are multiples of 3 *and* 4.

2 Evaluate these *without using a calculator*.
(a) $0{\cdot}5^2$ (b) 30^3 (c) $\sqrt{144}$ (d) $\sqrt[3]{1000}$ (e) $0{\cdot}2^3$ (f) $\sqrt{(10^2)}$
Use your calculator to check your answers.

3 Find the reciprocals of these numbers.
(a) 20 (b) $\frac{1}{5}$ (c) 400 (d) $\frac{2}{3}$ (e) 0·005 (f) 10^3

4 Look at this list of numbers.

1 5 12 13 25 37 47 64 80

From the list write down all the numbers which are
(a) prime, (b) multiples of 2 and 5, (c) square, (d) both square and cube.

5 Chris is thinking of an odd number which is prime and also a factor of 52.
What number is he thinking of?

Answers and hints ► page 100

Negative numbers and fractions

Negative numbers

Adding a negative number does the same as subtracting the equivalent positive number.
For example, $4 + {}^{-}2 = 4 - 2 = 2$ and ${}^{-}4 + {}^{-}2 = {}^{-}4 - 2 = {}^{-}6$.

Subtracting a negative number does the same as adding the equivalent positive number.
For example, $4 - {}^{-}2 = 4 + 2 = 6$ and ${}^{-}4 - {}^{-}2 = {}^{-}4 + 2 = {}^{-}2$.

The table shows the rules for the signs when you multiply or divide positive and negative numbers.
For example, $3 \times {}^{-}4 = {}^{-}12$ and ${}^{-}12 \div {}^{-}3 = 4$.

	× or ÷	Second number: pos	Second number: neg
First number	pos	pos	neg
	neg	neg	pos

1 Work out these.

(a) ${}^{-}8 + 5$ (b) $4 - {}^{-}7$ (c) ${}^{-}9 \times 4$ (d) ${}^{-}0{\cdot}5 \times {}^{-}6$

(e) $9 \div {}^{-}2$ (f) ${}^{-}20 \div (2 + {}^{-}6)$ (g) ${}^{-}\frac{1}{2}(14 - {}^{-}6)$ (h) $({}^{-}3{\cdot}6 - {}^{-}1{\cdot}2) \div 3$

2 The temperature is 9 °C.

(a) What will it be after it has fallen by 15 °C?

(b) If the temperature falls a further 4 °C, what will it be then?

3 Payments *into* bank accounts are called **credits** and payments *out* are **debits**.
Bev's account is overdrawn by £89·50 (she owes the bank £89·50).
How much does she have in her account after it is credited with £250 and £21·75 and debited for £85·34?

Fractions

$\frac{1}{2} = \frac{2}{4} = \frac{3}{6} = \frac{4}{8}$. They are called **equivalent fractions**.

These examples show how to add, subtract, multiply and divide fractions.

$\frac{2}{3} + \frac{1}{12} = \frac{2\times4}{3\times4} + \frac{1}{12} = \frac{8}{12} + \frac{1}{12} = \frac{9}{12} = \frac{3}{4}$ ← *Use equivalent fractions for addition and subtraction.*

$\frac{3}{8} - \frac{1}{3} = \frac{3\times3}{8\times3} - \frac{1\times8}{3\times8} = \frac{9}{24} - \frac{8}{24} = \frac{1}{24}$ ← *Notice how $\frac{9}{12}$ is simplified (put in its lowest terms).*

$\frac{2}{3} \times \frac{1}{4} = \frac{2\times1}{3\times4} = \frac{2}{12} = \frac{1}{6}$ ← *Multiply the top numbers together to get the top number. Multiply the bottom numbers together to get the bottom number.*

$\frac{3}{4} \div \frac{1}{8} = \frac{3}{4} \times 8 = \frac{24}{4} = 6$ ← *We need to find how many eighths are in $\frac{3}{4}$.*

$\frac{7}{8} \div \frac{2}{3} = \frac{7}{8} \times \frac{3}{2} = \frac{21}{16} = \frac{16+5}{16} = 1\frac{5}{16}$ ← *The 'top heavy' fraction $\frac{21}{16}$ is written as a mixed number $1\frac{5}{16}$.*

4 Work out these. Give your answers in their simplest form.

(a) $\frac{2}{3} + \frac{3}{5}$ (b) $1\frac{5}{8} - \frac{2}{3}$ (c) $\frac{2}{5} \times \frac{3}{4}$ (d) $\frac{9}{10} + \frac{3}{5} - \frac{3}{8}$

(e) $\frac{3}{4}$ of $1\frac{1}{2}$ (f) $\frac{7}{10} \div 5$ (g) $2\frac{1}{2} \div \frac{2}{3}$ (h) $\frac{1}{2}(\frac{1}{4} \times \frac{2}{5})$

5 In Ben's class three-fifths of the students come to school by car, a quarter come by bus and the remainder walk. What fraction of the class walks to school?

Answers and hints ► page 100

Getting rough answers; using a calculator

Rounding

The calculator display shows the result of dividing £33·50 by 7.
It is sensible to round this to the nearest penny (2 decimal places).

4.7857143

£4·7857143 = £4·79 (to the nearest penny)

Make sure you agree with the rounding of these numbers.

8762 = 9000 to the nearest thousand

745 = 750 to the nearest ten ⇐ *If the next figure is 5 or more, round up.*

8·275 = 8·3 to the nearest tenth (to 1 decimal place)

0·08314 = 0·083 to the nearest thousandth (to 3 decimal places)

Upper and lower bounds ► page 62

Significant figures

Look at the numbers 5491 and 0·005701.
For both numbers, the first *non-zero* significant figure is the 5.

Make sure you agree with these.

5491 = 5000 to 1 s.f. and 0·005701 = 0·006 to 1 s.f.

5491 = 5500 to 2 s.f. and 0·005701 = 0·0057 to 2 s.f.

5491 = 5490 to 3 s.f. and 0·005701 = 0·00570 to 3 s.f.

These two zeros are important because they keep the number the correct size.

This zero must be included because it is significant.

Using a calculator

Rounding to find rough answers is a useful skill. When using a calculator you should always check that your answer is about the right size.

Remember also that multiplying by a number less than one makes something smaller and dividing by a number less than one makes it larger.

Make sure you can use your calculator to do calculations like these:

(a) $50 - (8 \times 4)$ (b) $\dfrac{20}{2 \times 2{\cdot}5}$ (c) $100 - \dfrac{90}{30}$

You will need to use the memory and/or bracket keys on your calculator. Check by working them out in your head.

You should also be able to use your calculator efficiently to cope with 'constant' functions, for example $12 \times 1{\cdot}25$, $15 \times 1{\cdot}25$, $24 \times 1{\cdot}25$.

1 Round each of these as described in the brackets.

(a) 4·085 [to 2 d.p.] (b) 6·049 [to 1 d.p.] (c) 0·0097 [to 3 d.p.]

2 Round off:

(a) 3755 to 2 s.f. (b) 0·0783 to 2 s.f. (c) 0·0704 to 2 s.f.
(d) 3·042 to 2 s.f. (e) 329·7 to 3 s.f. (f) 0·987 to 1 s.f.

3 Show clearly how you would find an **estimate** for this calculation: $680^2 \div (53 \times 0{\cdot}92)$

4 Without using a calculator, explain why $\dfrac{15{\cdot}52 \times 0{\cdot}913}{102{\cdot}8 \times 0{\cdot}0314}$ is about 5. MEG (SMP)

5 Do these on your calculator. Give each answer correct to 2 significant figures.

(a) $\dfrac{58{\cdot}12}{48{\cdot}2 + 66{\cdot}4}$ (b) $\dfrac{8{\cdot}5(10{\cdot}8 - 3{\cdot}2)}{0{\cdot}048^2}$ (c) $\dfrac{0{\cdot}36 \times \sqrt{320}}{8{\cdot}71 - 3{\cdot}09}$ (d) $\dfrac{(4{\cdot}2 + 3{\cdot}6)^2}{\sqrt{33{\cdot}7}}$

Check the size of each answer by rounding.

6 Sylvia and Graham are using the formula $v^2 = u^2 - 2gh$.

(a) Sylvia has got to the stage $v = \sqrt{(18{\cdot}6^2 - 2 \times 9{\cdot}8 \times 7{\cdot}6)}$.

(i) Work this out on your calculator. Write down the full calculator display.

(ii) Write down your answer to a reasonable degree of accuracy.

(b) Graham has got to the stage $h = \dfrac{29{\cdot}5 - 19{\cdot}7^2}{2 \times 9{\cdot}8}$.

He works out the answer as 850·45, but his answer is wrong.
Write down a calculation he could have done in his head to check his answer. MEG (SMP)

7 Colin buys 28·4 litres of petrol at 52·5p a litre.

The price at another garage is 46·2p a litre.

How much more petrol can Colin buy for the same money if he uses this garage? MEG/ULEAC (SMP)

8 (a) Leza bought a CD player.
She paid £28·00 deposit and 9 monthly payments of £8·40.
How much did she pay altogether?

(b) Sandra also bought a CD player.
She paid 21 weekly payments of £5·20 but no deposit.
How much more did she pay than Leza? NICCEA

9 (a) Tim changed some money into dollars for a trip abroad. The rate of exchange was \$1·578 to £1. How much did Tim expect to receive in dollars for £800?

(b) In fact, the bank said he must purchase \$1300 at this rate of exchange and also pay £5 commission. How much did Tim have to pay altogether? MEG/ULEAC (SMP)

10 An encyclopedia has a total thickness of 67·5 mm.
The front and back covers are each 4·0 mm thick.
The book has 1146 pages.

Without using a calculator, or long division, estimate the thickness in millimetres of a single sheet of paper.
(Remember that each sheet of paper has one page on each side.)

Show your method of estimation. MEG (SMP)

Answers and hints ► page 101

Fractions, decimals and percentages

Converting between fractions, decimals and percentages

Make sure you understand these examples.

$9{\cdot}5\% = \frac{9{\cdot}5}{100} = 0{\cdot}095$ $\quad$ $\frac{1}{20} = 0{\cdot}05 = 0{\cdot}05 \times 100\% = 5\%$ or $\frac{1}{20} = \frac{5}{100} = 5\%$

$0{\cdot}85 = 0{\cdot}85 \times 100\% = 85\%$ $\quad$ $1{\cdot}27 = 1{\cdot}27 \times 100\% = 127\%$

Expressing one quantity as a percentage of another

If 15 out of 32 students in a class are girls, the **percentage** of girls is given by

$\frac{15}{32} \times 100 = 46{\cdot}875 = 47$ (to 2 s.f.).

Finding a value after a percentage increase or decrease

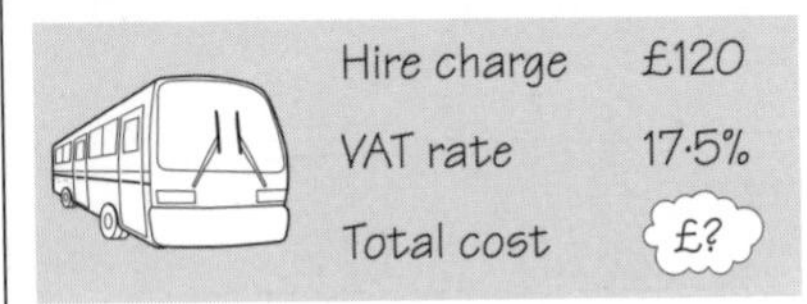

Total cost = 117·5% of hire charge (that is 100% + 17·5%)
= £120 × 1·175 *The multiplying factor is 1·175.*
= £141

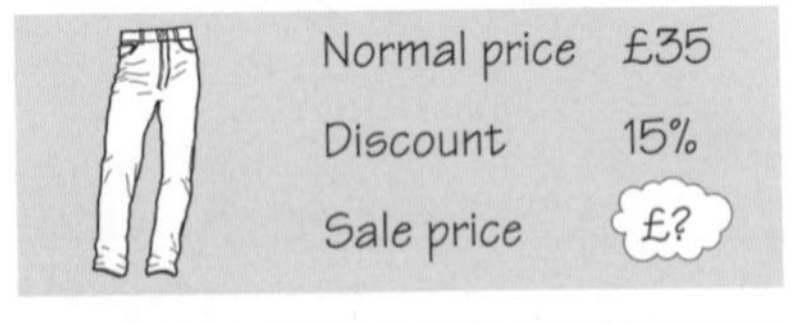

Sale price = 85% of normal price (that is 100% – 15%)
= £35 × 0·85 *The multiplying factor is 0·85.*
= £29·75

1 Write these as decimals.
(a) 84% (b) $\frac{9}{10}$ (c) 2% (d) $\frac{3}{16}$ (e) 12·5%

2 Write these as percentages.
(a) 0·7 (b) $\frac{5}{8}$ (c) 0·725 (d) $\frac{13}{20}$ (e) 2·4

3 Write each of these as a fraction in its simplest form.
(a) 75% (b) 0·12 (c) 17·5% (d) 0·05 (e) 6·5%

4 Work out each of these.
(a) 20% of 65 kg (b) 5% of £15 (c) 30% of £1·80 (d) 175% of 4·8 m

5 (a) Change $\frac{11}{17}$ into a decimal.
Give your answer correct to 3 decimal places.

(b) Place the following numbers in order of size, starting with the smallest.

$\frac{11}{17}$ 65% $\frac{3}{5}$ 0·63

MEG (SMP)

6 Show by calculation which is the larger of $\frac{4}{11}$ and $\frac{3}{8}$.

MEG (SMP)

7 The bill for servicing Peter's car is £79·25 + VAT.
What is the total cost if the VAT rate is 17·5%?
Give your answer to the nearest penny.

8 Christopher wants to buy a portable CD–cassette player with a marked price of £160.
The shopkeeper offers him the following hire purchase terms:
'a deposit of 25% of the marked price followed by 12 monthly payments of £14·40'.
(a) How much is the deposit required?
(b) How much more than the marked price would Christopher have to pay by hire purchase?

WJEC

9 A holiday excursion takes people on trips into the desert.
People can travel by jeep or camel.
This table shows how people in different age groups chose to travel by each method in one month.

	Jeep	Camel
Under 30	71	69
Over 30	39	21

(a) Copy and complete this table to show the percentage of each age group that went by jeep and the percentage of each age group that went on camels.

	Jeep	Camel
Under 30	50·7%	..........%
Over 30	..........%	..........%

(b) This table gives the data in a different percentage form.
Write down in your own words what the shaded percentage represents.

	Jeep	Camel
Under 30	64·5%	76·7%
Over 30	35·5%	23·3%
	100%	100%

MEG (SMP)

10 In the 1996/97 tax year, the basic personal tax-free allowance was £3765.
The table shows the income tax rates *on taxable income* that applied over this period.
Mary is single and earned £32 600 before deductions.
How much income tax did she pay?

Bands £	**Rate** %
1–3900	20
3901–25 500	24
Over 25 500	40

11 Copy and complete this table.

Membership of a National Charity

	Men	Women	Total
Live in Scotland	415 678	267 350	
Live elsewhere	73 897	105 951	
Total			

(a) What percentage of the men live in Scotland?
(b) What percentage of the membership are women?

MEG/ULEAC (SMP)

Answers and hints ► page 101

Percentages

Finding a percentage increase or decrease

Example

The price of a television is reduced from £220 to £176 in a sale.
What is the percentage reduction?

There are two ways of doing this:

Work out the increase or decrease.

Decrease $= £220 - £176 = £44$

Percentage decrease $= \dfrac{\text{decrease}}{\text{old price}} \times 100$

$= \frac{44}{220} \times 100 = 20$

or

Write the fraction $\dfrac{\text{new price}}{\text{old price}}$ and change it to a percentage.

$\frac{176}{220} = 0{\cdot}8 = 80\%$

The new price is 80% of the old price.

The percentage decrease is $100 - 80 = 20$.

Compound percentages

Example

£500 is invested for two years.

In the first year the interest rate is 5%. In the second year it is 4%.

How much money is in the account after 2 years?

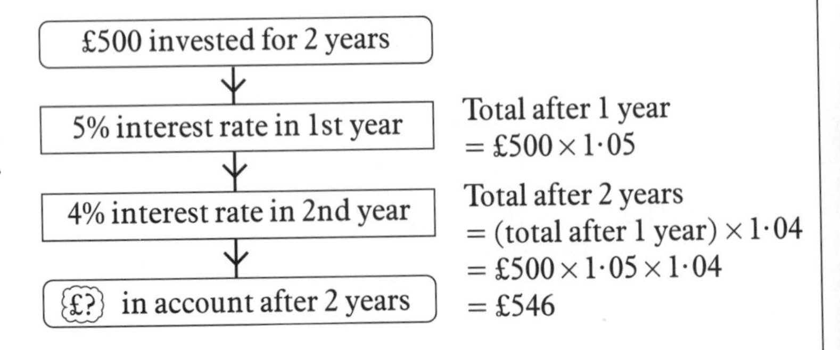

Percentages backwards

Example

A bicycle costs £140 after a 12% increase in price.
How much did it cost before the increase?

Old price $\times 1{\cdot}12 = £140$

So the old price $= £140 \div 1{\cdot}12$ *Divide both sides of the equation by 1·12.*

$= £125$

1 The price of a gallon of petrol is increased from £2·25 to £2·39.
What is the percentage increase?

2 On *Liftoff* flights the fare from London to New York is £310.
From March 31st the fares will be increased by 13%.
A discount of 4·5% will be offered on night flights.

How much will a ticket for a night flight from London to New York cost after March 31st?

MEG (SMP)

3 Mike is offered a job as a trainee accountant.
He can expect an 8% increase in his salary when he passes his first year exams.
He can also expect a further 15% increase when he passes the second year exams.
What overall percentage increase can he expect if he passes both examinations? MEG (SMP)

4 In 1990, a charity sold $2\frac{1}{4}$ million lottery tickets at 25p each.
80% of the money obtained was kept by the charity.
(a) Calculate the amount of money kept by the charity.

In 1991, the price of a lottery ticket fell by 20%.
Sales of lottery tickets increased by 20%.
80% of the money obtained was kept by the charity.
(b) Calculate the percentage change in the amount of money kept by the charity. ULEAC

5 The manager of a carpet shop has made a mistake in writing an advertisement.
The prices are correct, but the 'percentage off' is NOT correct.
(a) Find, correct to one decimal place, the true 'percentage off'.
(b) In another shop, where a genuine reduction of 20% had been made, the sale price of a carpet was also £6·39.
What was the original price in this shop? NICCEA

6 To encourage careful driving, insurance companies offer discounts to drivers who have no accidents.
Lisa has 3 years' no claim discount.
She pays a premium of £268·80.
What would her premium have been without her discount?

1 year's no claims discount	15%
2 years' no claims discount	25%
3 years' no claims discount	35%
4 years' no claims discount	50%

7 It is estimated that the number of rabbits on Warren Island is decreasing at the rate of 12% per year.
In 1993 the number of rabbits was 308.
(a) How many rabbits were there in 1994?
(b) How many rabbits were there in 1992?
(c) In which year will there first be less than 180 rabbits? MEG (SMP)

8 Bernard has to exchange money for a touring holiday in France and Italy.
The bank **first** calculates its commission at the rate of 1·5%.
The balance is then exchanged into the appropriate currency.
(a) How many francs will Bernard receive for £300 after allowing for commission?
(b) How much would it cost Bernard to buy 1·5 million lire? MEG (SMP)

Currency	Rate per £
franc	8·40
lire	2320

Answers and hints ► page 102

Ratio and proportion

Simplest form

The **ratio** of grey squares to white squares in each of these diagrams is 1 to 2 or 1:2.

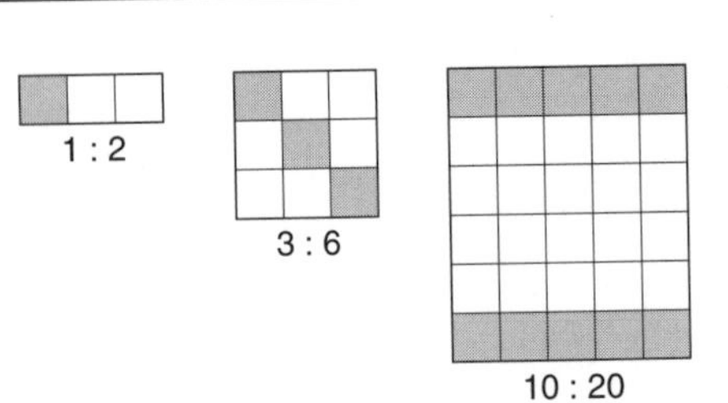

1:2 is the **simplest form** of the ratios 3:6 and 10:20.
We say that these three ratios are **equivalent**.

Sharing in a given ratio

To share £200 in the ratio 2:3, divide £200 into 2 + 3 = 5 equal parts.

$\frac{1}{5}$ of £200 = £40.

So the shares of the money will be $2 \times £40 = £80$ and $3 \times £40 = £120$.

Remember to check by adding the shares: 80 + 120 = 200.

Proportional quantities and their graphs

Types of proportionality ► page 12

The table shows the amount of calcium in various amounts of milk.
If this data is plotted, the graph is a straight line passing through (0, 0).
Any two quantities which are related in this way are said to be **proportional**.

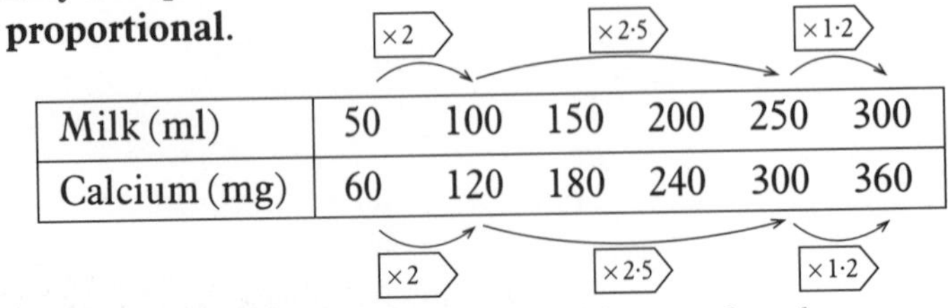

Milk (ml)	50	100	150	200	250	300
Calcium (mg)	60	120	180	240	300	360

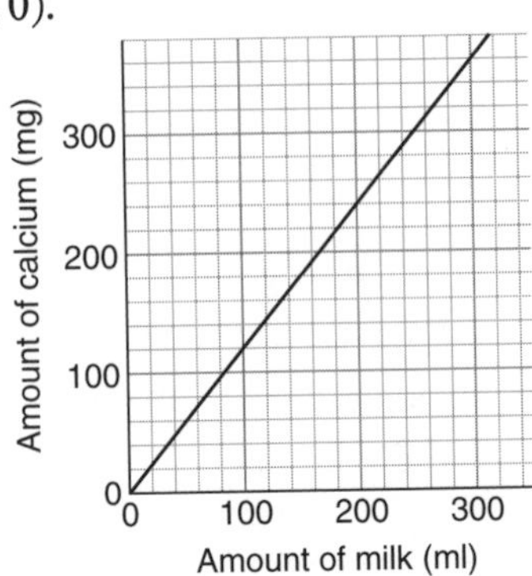

Notice how the quantities in the table are related.

Suppose you wanted to find out how much calcium there is in half a pint of milk (1 pint = 568 millilitres).

You could either read the answer from the graph or *calculate* using one of these two methods.

Multiplier method for calculating proportion

Multiplying one quantity by a number means multiplying the other quantity by the same number.

50 ml	×?	284 ml	*Multiplier*
60 mg	×?	? mg	

Divide 284 by 50 to get the **multiplier** 5·68.

So the amount of calcium in 284 ml = $60 \times 5{\cdot}68$ mg = 341 mg (to the nearest mg).

Unitary method for calculating proportion

50 ml contains 60 mg.

÷50 ÷50 *Divide by 50 to find the amount of calcium in 1 ml.*

1 ml contains $\frac{60}{50}$ mg.

×284 ×284 *Multiply by 284 to find the amount of calcium in 284 ml.*

284 ml contains $\frac{60}{50} \times 284$ mg = 341 mg (to the nearest mg).

1 Write these ratios in their simplest form.

(a) 24 : 18 (b) $3\frac{1}{3}$: 10 (c) 35p : £1 (d) 1 hour : 12 minutes

2 A chain saw needs a mixture of oil and petrol in the ratio 1 : 25.
Peter has measured out 800 ml of petrol.
How much oil must he add to it?

3 Anne is 12 years old and her brother Bobby is 8 years old.
They have £2 pocket money altogether.
The money is shared in the ratio of their ages.

(a) Calculate the amount each receives.

Two years ago Bobby received 30p pocket money.
Their pocket money was in the same ratio as their ages then.

(b) Calculate how much money Anne received then. ULEAC

4 The cost of electricity is proportional to the wattage.
The electricity for an ordinary 60 watt light bulb costs about 36p a week.

How much would a low energy 15 watt bulb cost per week? MEG (SMP)

Ordinary 60 watt light bulb

15 watt low energy light bulb

5 An electrician charges a £25 call-out fee and £20 per hour.

(a) How much would he charge if he took $2\frac{1}{2}$ hours to mend a washing machine?

(b) Is the amount charged directly proportional to the time taken?
Explain your answer.

6 An alloy is made of copper, tin and phosphorous in the ratio 185 : 14 : 1.

(a) Find the amount of copper in 120 kg of the alloy.

(b) Calculate the amount of alloy that has 4·9 kg of tin in it. WJEC

7 A 'perf' is the number of perforations in 2 cm along one edge of a stamp.

A stamp has 27 perforations along an edge which is 3·6 cm long.
What is the 'perf' for this stamp? MEG (SMP)

Not to scale

8 A gas cylinder contains gas at a temperature of 288 K and a pressure of $1{\cdot}80 \times 10^5$ N/m^2.
What will the pressure be if the gas is heated to 303 K, assuming that the pressure is proportional to the temperature?

Answers and hints ► page 103

Types of proportionality

Direct and inverse proportionality

When two variables p and q are **directly proportional** ($p \propto q$), the graph looks like this.

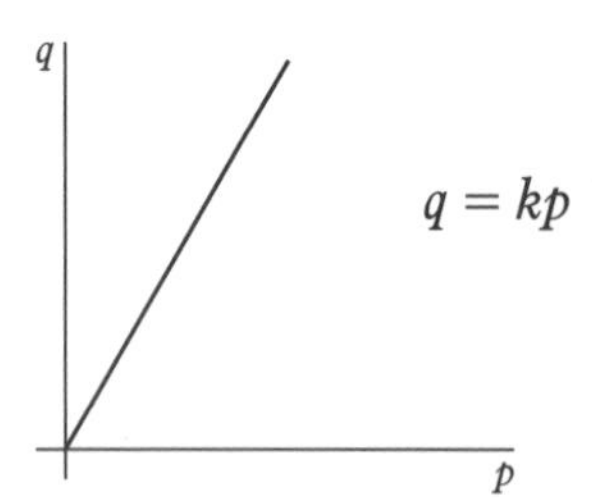

The multiplier for q = the multiplier for p.

The ratio $\frac{q}{p}$ is constant and equal to the gradient of the line.

When two variables p and q are **inversely proportional** ($q \propto \frac{1}{p}$), the graph looks like this.

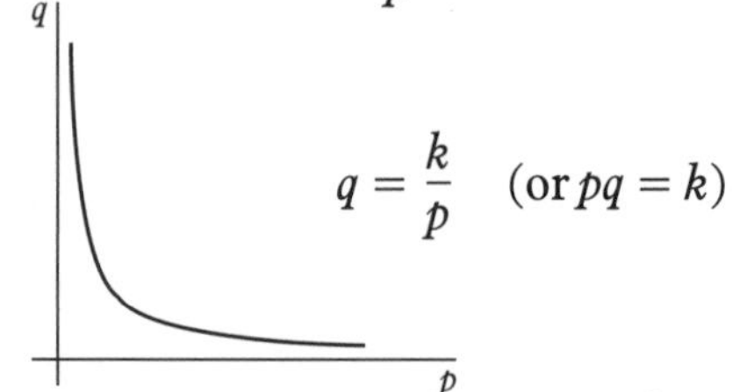

The multiplier for $q = \frac{1}{(\text{multiplier for } p)}$.

The product pq is constant.

Proportional quantities and their graphs ► page 10

Example

Complete this table given that p is inversely proportional to q.

p	3	6	60	15
q	20			

Start by working out the multipliers for successive values of p.

Then use the corresponding multipliers to work out the values of q.

Multipliers for p: $\times 2$, $\times 10$, $\times \frac{1}{4}$

p	3	6	60	15
q	20	10	1	4

Multipliers for q: $\times \frac{1}{2}$, $\times \frac{1}{10}$, $\times 4$

Other types of proportionality

There are three more types of proportion which you need to know. Notice the effect on the values in the tables.

$q \propto p^2$ $\quad q = kp^2$ $\quad$ multiplier for q = (multiplier for p)^2

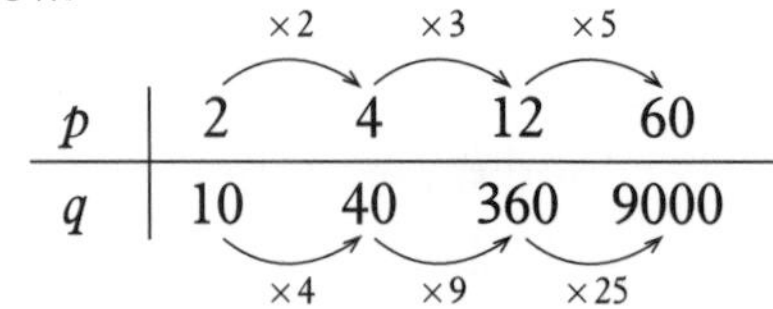

Multipliers for p: $\times 2$, $\times 3$, $\times 5$

p	2	4	12	60
q	10	40	360	9000

Multipliers for q: $\times 4$, $\times 9$, $\times 25$

$q \propto p^3$ $\quad q = kp^3$ $\quad$ multiplier for q = (multiplier for p)^3

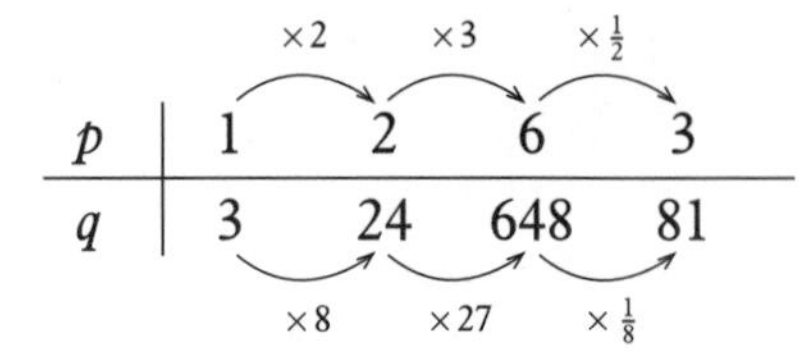

Multipliers for p: $\times 2$, $\times 3$, $\times \frac{1}{2}$

p	1	2	6	3
q	3	24	648	81

Multipliers for q: $\times 8$, $\times 27$, $\times \frac{1}{8}$

$q \propto \frac{1}{p^2}$ $\quad q = \frac{k}{p^2}$ $\quad$ multiplier for $q = \frac{1}{(\text{multiplier for } p)^2}$

Multipliers for p: $\times 2$, $\times 10$, $\times \frac{1}{5}$

p	1	2	20	4
q	2	$\frac{1}{2}$	$\frac{1}{200}$	$\frac{1}{8}$

Multipliers for q: $\times \frac{1}{4}$, $\times \frac{1}{100}$, $\times 25$

You can find the value of k by substituting a pair of known values in the relevant equation: in the last example, q is 2 when p is 1, so $k = qp^2 = 2$.

Fitting functions to data ► page 40

1 Copy and complete the table of values on the assumption that
(a) y is proportional to x^2, (b) y is proportional to $\frac{1}{x}$.

x	6	24	48	120
y		100		

2 y is inversely proportional to the square of x, and $y = 10$ when $x = 2$.
(a) Write down an equation connecting y and x.
(b) Calculate (i) y when $x = 4$, (ii) x when $y = 40$.

3 The Highway Code gives this table for the braking distance of cars.

Speed in miles per hour (x)	30	40	50	60	70
Braking distance in feet (y)	45	80	125	180	245

(a) y is proportional to x^2. Write an equation connecting y and x^2.
(b) Use your equation to find the braking distance in feet for a speed of 75 m.p.h.
(c) What is the speed in miles per hour when the braking distance is 400 feet?

4 The density of a gas is inversely proportional to its volume.
What happens to the density when the volume is increased by a factor of 1·5?

5 The energy stored in a battery is proportional to the square of the diameter of the battery, for batteries of the same height.
One battery has a diameter of 2·5 cm and stores 1·6 units of energy.
Another has a diameter of 1·5 cm.
Calculate the energy stored in the second battery.

ULEAC

6 (a) Copy and complete the following table if $d \propto t^2$ and $F \propto \frac{1}{d}$.

t	3	30	15	5	50	100
d	180					
F	1000					

(b) Write down equations connecting (i) d and t, (ii) F and d.

7 The frequency of sound is inversely proportional to the wavelength.
The lowest audible sound has a frequency of 20 Hertz and a wavelength of 16·5 metres.
(a) A sound has wavelength 1 metre.
What is its frequency?
(b) The highest audible sound has a frequency of 15 000 Hertz.
What is its wavelength?

MEG (SMP)

8 In the table Q is proportional to the cube of P.
Calculate s and t.

P	0·8	t	6
Q	s	13·5	108

Answers and hints ► page 104

Indices

Powers

A number such as 10000 can be written as a **power** of ten: $10000 = 10 \times 10 \times 10 \times 10 = 10^4$

If we continue the pattern of powers to the right of the decimal point, the powers become negative.

ten thousands	thousands	hundreds	tens	ones (or units)	tenths	hundredths	thousandths	ten thousandths
10^4	10^3	10^2	10^1	10^0	10^{-1}	10^{-2}	10^{-3}	10^{-4}

We use 10^{-1} to mean **1 tenth,**
10^{-2} to mean **1 hundredth,** and so on.

Notice that 10^0 is 1.

Standard form

Powers of ten are useful for expressing large and small numbers. For example:

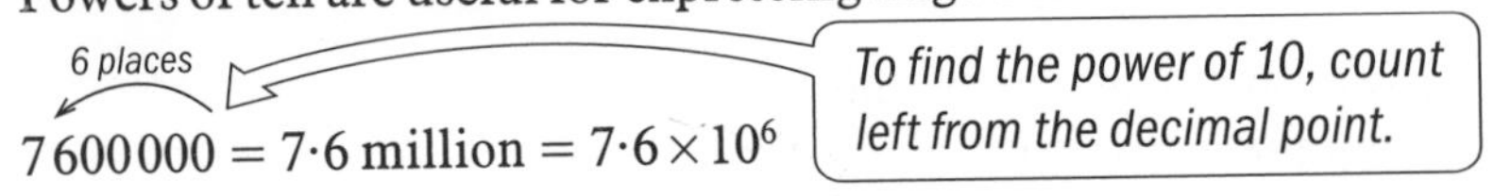

$7600000 = 7{\cdot}6 \text{ million} = 7{\cdot}6 \times 10^6$

$7{\cdot}6 \times 10^6$ is called the **standard form** of 7600000.
Note that the first part has to be a number between 1 and 10.

Make sure you can use the exponent key on your calculator to handle numbers in standard form.

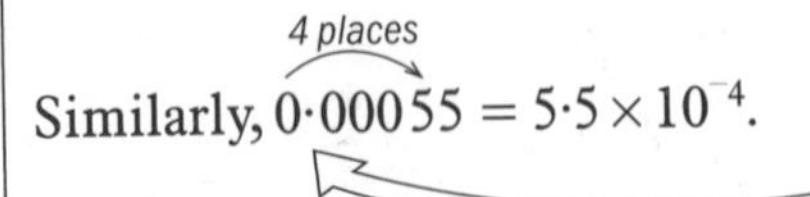

Similarly, $0{\cdot}00055 = 5{\cdot}5 \times 10^{-4}$.

To find the power of 10, count right from the decimal point.

Simplifying expressions involving indices

The rules of indices ► page 20

*To multiply powers of one number together you **add** the indices.*

$5^2 \times 5^4 = (5 \times 5) \times (5 \times 5 \times 5 \times 5) = 5^6$

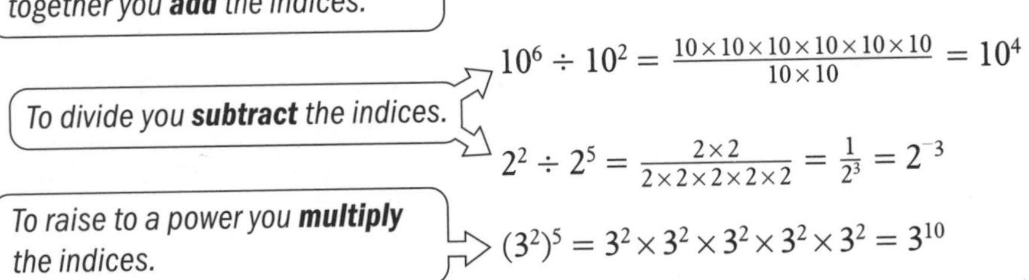

$10^6 \div 10^2 = \frac{10 \times 10 \times 10 \times 10 \times 10 \times 10}{10 \times 10} = 10^4$

$2^2 \div 2^5 = \frac{2 \times 2}{2 \times 2 \times 2 \times 2 \times 2} = \frac{1}{2^3} = 2^{-3}$

$(3^2)^5 = 3^2 \times 3^2 \times 3^2 \times 3^2 \times 3^2 = 3^{10}$

Fractional indices

According to the rules of indices: $9^{\frac{1}{2}} \times 9^{\frac{1}{2}} = 9^{(\frac{1}{2} + \frac{1}{2})} = 9^1$

This means that a power of a half must represent a square root: $9^{\frac{1}{2}} = \sqrt{9} = 3$

Similarly a power of a third represents a cube root: $27^{\frac{1}{3}} = \sqrt[3]{27} = 3$

1 Convert these numbers to ordinary form.

(a) $6{\cdot}6 \times 10^3$ (b) 7×10^{-5} (c) $4{\cdot}83 \times 10^{-2}$ (d) $16^{\frac{1}{2}}$ (e) $1000^{-\frac{1}{3}}$

2 Write these numbers in standard form.

(a) 523000 (b) 0·000046 (c) 0·01007 (d) 78110000 (e) $\frac{1}{10000}$

3 Simplify these. Write your answers using index notation.

(a) $5^4 \times 5^3$ (b) $3^7 \div 3^{-3}$ (c) $7^{\frac{1}{3}} \times 7^3 \times 7^{-2}$ (d) $(8^4)^{\frac{1}{2}}$

(e) $27 \div 3^5$ (f) $11^{-4} \div 11^0$ (g) $2^{-5} \div 2^{-1}$ (h) $4^{\frac{1}{2}} \times 16^{\frac{1}{2}}$

4 Express the following as powers of 2.

(a) 8 (b) $\frac{1}{4}$ (c) 512 (d) $\frac{1}{64}$ (e) $\frac{1}{\sqrt{2}}$

5 Express the following in index form using prime numbers.

(a) 49 (b) 27 (c) $\frac{1}{8}$ (d) $\sqrt{27}$ (e) $\sqrt[3]{4}$ (f) $\frac{1}{\sqrt{625}}$

6 Use the power key on your calculator to evaluate these.

(a) 5^{10} (b) 8^7 (c) $50^{\frac{1}{3}}$ (d) $0{\cdot}8^{-4}$ (e) $120^{-\frac{2}{3}}$

7 Calculate these to 3 s.f., giving your answers in standard form.

(a) $\dfrac{6{\cdot}31 \times 10^8}{4{\cdot}52 \times 10^{-5}}$ (b) $\dfrac{6{\cdot}31 \times 10^{12}}{\sqrt{(5{\cdot}56 \times 10^6)}}$

8 A 'googol' is the mathematical name for the number 1 followed by one hundred zeros.
Write the number '15 googols' in standard form. MEG (SMP)

9 New York produces more waste per person than any other city in the world.
It produces an average of 1·8 kg per day for each person.
The population is about $1{\cdot}4 \times 10^7$ people.

About how many tonnes of waste each year is this for the whole city?
Give your answer in standard form.
[1 tonne = 10^3 kg]

MEG/ULEAC (SMP)

10 A light year is the distance travelled by light in one year, approximately 9 500 000 000 000 km.

(a) Express the number of kilometres in one light year in standard form.

The star, Proxima Centauri, is 4·3 light years away from the Earth.

(b) Calculate the distance, in km, from Proxima Centauri to the Earth.
Give your answer in standard form.

The Sun is approximately 150 000 000 km from the Earth.

(c) How many times further from the Earth is Proxima Centauri than the Sun? ULEAC

11 (a) Write the number 0·000 345 in standard form.

(b) If $A = 5 \times 10^{-8}$, write the value of $\dfrac{1}{A}$ in standard form.

(c) The population density of a country is the number of people per square mile.
Sudan has a population of $2{\cdot}52 \times 10^7$ and an area of $9{\cdot}67 \times 10^5$ square miles.
Calculate the population density of Sudan. MEG (SMP)

Answers and hints ► page 105

Rational and irrational numbers

Types of decimal

$$8\overline{)3{\cdot}0000}\quad\text{quotient } 0{\cdot}3750$$

$\frac{3}{8} = 0{\cdot}375$, a **terminating** decimal

$$7\overline{)3{\cdot}000000000}\quad\text{quotient } 0{\cdot}428571428\ldots$$

$\frac{3}{7} = 0{\cdot}\dot{4}2857\dot{1}$, a **recurring** decimal

The numbers within the dots recur.

This decimal must recur after at most 6 digits because there are only 6 possible remainders.

Any fraction will give either a terminating or recurring decimal.

Rational and irrational numbers

A **rational** number is a number which can be written in the form $\frac{a}{b}$ (where a and b are integers and b is non-zero).

An **irrational** number is one which cannot be written as a fraction: two examples are π and $\sqrt{2}$.

Changing recurring decimals to fractions

Example Change $0{\cdot}7\dot{3}\dot{1}$ into a fraction.

Let f be the fraction which is equivalent to $0{\cdot}7\dot{3}\dot{1}$.

$$f = 0{\cdot}731313131\ldots$$

$$100f = 73{\cdot}131313131\ldots$$ ← *Multiply both sides of the equation by 100.*

$$99f = 72{\cdot}4$$ ← *Subtract the first equation from the second. (All the 'three ones' go.)*

$$f = \frac{72{\cdot}4}{99}$$ ← *Divide both sides by 99.*

$$f = \frac{724}{990} = \frac{362}{495}$$ ← *Multiply by 10 and divide by 2 to simplify the fraction.*

Check your answer by using a calculator to change the fraction back to a decimal.

Surds

A **surd** is an irrational root. You need to be able to manipulate surds.

A helpful rule is $\sqrt{a} \times \sqrt{b} = \sqrt{(ab)}$.

This is useful in simplifying expressions. For example:

$$\sqrt{500} = \sqrt{(100 \times 5)} = \sqrt{100} \times \sqrt{5} = 10\sqrt{5}$$

$$\frac{6}{\sqrt{2}} = \frac{6\sqrt{2}}{\sqrt{2} \times \sqrt{2}} = \frac{6\sqrt{2}}{\sqrt{4}} = \frac{6\sqrt{2}}{2} = 3\sqrt{2}$$ ← *Start by multiplying the top and bottom of the fraction by $\sqrt{2}$.*

$$\begin{aligned}(\sqrt{2} + 1)(3 - 2\sqrt{2}) &= 3\sqrt{2} - (\sqrt{2} \times 2\sqrt{2}) + 3 - 2\sqrt{2}\\ &= 3\sqrt{2} - 4 + 3 - 2\sqrt{2}\\ &= \sqrt{2} - 1\end{aligned}$$

Expanding double brackets ► page 31

1 Change these recurring decimals to fractions.

(a) $0{\cdot}\dot{3}\dot{6}$ (b) $1{\cdot}\dot{2}$ (c) $0{\cdot}0\dot{4}\dot{7}$ (d) $0{\cdot}5\dot{6}\dot{1}$ (e) $0{\cdot}6\dot{3}1\dot{7}$

2 An approximate value of $\sqrt{3}$ is 1·73.
Use this to find approximate values for (a) $\sqrt{300}$ (b) $\sqrt{27}$

3 Simplify each expression as much as possible, removing surds from the denominator.

(a) $\sqrt{8} \times \sqrt{2}$ (b) $\dfrac{6\sqrt{2}}{\sqrt{6}}$ (c) $(\sqrt{3} + \sqrt{2})^2$ (d) $\sqrt{48} + \sqrt{3}$

(e) $\dfrac{\sqrt{7000}}{\sqrt{700}}$ (f) $(\sqrt{12} - \sqrt{3})^2$ (g) $(\sqrt{5} + 2)(2 - \sqrt{5})$ (h) $\dfrac{4}{\sqrt{2}} + \dfrac{3}{\sqrt{2}}$

4 (a) $\sqrt{3}$ is an irrational number. Consider the true statement
$$\sqrt{12} = \sqrt{(4 \times 3)} = \sqrt{4} \times \sqrt{3} = 2\sqrt{3}$$
Does this show $\sqrt{12}$ to be rational or irrational? Explain your reasoning.

(b) $\sqrt{5}$ is irrational. Consider the true statement
$$\sqrt{36} = \sqrt{(5 \times 7{\cdot}2)} = \sqrt{5} \times \sqrt{7{\cdot}2}$$
What can be said about $\sqrt{7{\cdot}2}$? Explain your reasoning. MEG (SMP)

5 In this question $p = 3 + \sqrt{5}$ and $q = 3 - \sqrt{5}$.
Say whether each of these is rational or irrational.

(a) p (b) $p + q$ (c) $p - q$ (d) $p \times q$ MEG (SMP)

6 (a) A mathematics student attempted to define an 'irrational number' as follows:
'An irrational number is a number which, in its decimal form, goes on and on.'
Give an example to show that this definition is not correct.

(b) Which of the following numbers are rational and which irrational?

$\sqrt{(4\frac{1}{4})}$ $\sqrt{(6\frac{1}{4})}$ $\frac{1}{3} + \sqrt{3}$ $(\frac{1}{3}\sqrt{3})^2$

Express each of the rational numbers in the form $\dfrac{p}{q}$ where p and q are integers. MEG

7 Write these numbers in a suitable form to show which are rational
and which irrational, and state which is which.

(a) $0{\cdot}\dot{3}$ (b) $4^{\frac{1}{2}}$ (c) $8^{-\frac{1}{2}}$

(d) 3·142 (e) $(\sqrt{2})^{-4}$ (f) $(1 + \sqrt{5})(1 - \sqrt{5})$ MEG/ULEAC (SMP)

8 *Do not use a calculator in this question.*
$p = 3$ $q = {}^{-}4$

(a) Work out the following, showing your working.
Leave your answers as integers, fractions or irrational numbers.
(i) $p^2 - q^2$ (ii) $p^{\frac{1}{3}}$ (iii) q^{-2}

(b) One of the numbers in part (a) is irrational.
(i) State which one.
(ii) Explain why the other two numbers are rational.

(c) Simplify, as far as possible, $(5 - \sqrt{3})^2$. MEG (SMP)

Answers and hints ► page 105

Mixed number

1

		1		Sum = 1	$= 1^3$
	3		5	Sum = 8	$= 2^3$
7		9	11	Sum = 27	$= 3^3$
…	…	…	…	Sum = …	= …

(a) Copy and complete the fourth line of the number pattern above.

(b) Calculate which line will have a sum equal to 729. ULEAC

2 The table shows the polpulation of the continents.

Continent	Population
Africa	$5{\cdot}11 \times 10^8$
Asia	$2{\cdot}69 \times 10^9$
North America	$2{\cdot}52 \times 10^8$
South America	$3{\cdot}93 \times 10^8$
Europe	$6{\cdot}95 \times 10^8$
Australasia	$2{\cdot}40 \times 10^7$

(a) Which continent has the largest population?

(b) (i) What is the difference in population between North and South America?

(ii) Calculate the total population of Asia and Australasia, giving your answer in standard form.

(c) The population density of a continent is the number of people per square kilometre.
The continent of North America covers an area of $1{\cdot}93 \times 10^7$ km^2.
Calculate the population density of North America in people per square kilometre.

(d) The population of Africa is given to the nearest million.
Write down the lower and upper bounds of this population. MEG (SMP)

3 The volume, V m^3, of a certain mass of air is inversely proportional to the pressure, P pascals. If the volume was $7{\cdot}3 \times 10^{-4}$ m^3 when the pressure was $1{\cdot}01 \times 10^5$ Pa (atmospheric pressure), what will the volume be when the pressure is $1{\cdot}89 \times 10^5$ Pa?

4

Recipe for Shortcake Biscuits

$3\frac{3}{4}$ pounds of flour

$2\frac{1}{2}$ pounds of margarine

$1\frac{1}{4}$ pounds of sugar

Bakery students used this recipe to make a large number of biscuits.

(a) What fraction of the total weight of the ingredients is sugar?
Give your answer as a fraction in its lowest terms.

(b) John used $1\frac{1}{2}$ pounds of sugar instead of $1\frac{1}{4}$ pounds.
Calculate the percentage error in the amount of sugar used. MEG/ULEAC (SMP)

5 In the first 50 weeks of 1992, Barclays Bank dispensed £8·7 billion through its cash machines in 187 million transactions.

(a) Write as numbers in standard form:
(i) 8·7 billion (ii) 187 million (1 billion = 1 thousand million)

(b) What was the average amount of cash dispensed at each transaction?

(c) Show your working for a rough estimate to (b) to check that your answer is of the right order of magnitude.

On 19 December 1992 the cheque authorisation service authorised 34 200 cheques worth £4·37 million in total. Assume these numbers are both given to 3 significant figures.

(d) State the upper and lower bounds of
(i) the number of cheques, (ii) their total value.

Upper and lower bounds ► page 62

On the last Saturday in November in 1992, Barclays Bank machines dispensed £42·5 million. This was an increase of 2·5% from 1991.

(e) How much was dispensed on that Saturday in 1991?

MEG/ULEAC (SMP)

6 The average mass of 1 cubic metre of the Earth is $5{\cdot}51 \times 10^3$ kg.

(a) Write $5{\cdot}51 \times 10^3$ as an integer.

The volume of the Earth is $1{\cdot}08 \times 10^{21}$ m^3.

(b) Calculate the mass of the Earth in kg. Give your answer in standard form.

(c) Write your answer to part (b) in tonnes. Give your answer in standard form.

The mass of the Sun is 2×10^{27} tonnes.

(d) Find the mass of the Earth as a percentage of the mass of the Sun.

MEG/ULEAC (SMP)

7 A box of 240 tea bags contains six foil packs, each holding 40 bags. In an investigation at school Jenny found that each of the six foil packs weighed 135 grams correct to the nearest 5 grams.

(a) Obtain lower and upper bounds for the total weight of the six foil packs.

(b) Why would you not expect the total weight to be as low as this lower bound?

(c) Jenny was told that the scales she was using were faulty.
The true weight could be up to 5% above or below the indicated weight.
Investigate whether she has evidence that the label on the box (net weight 850 grams) is misleading.

MEG (SMP)

Answers and hints ► page 106

ALGEBRA

Simplifying and substituting

Simplifying by combining like terms

Make sure you can see that terms have been combined correctly in these.

$3a - a^2 - 7a + 4a^2 = 3a^2 - 4a$

$a^2 - ab - 2b^2 + 4ab + 2a^2 + b^2 = 3a^2 - b^2 + 3ab$

$^-4a + 3a^2$ is also correct but it is usual to put square terms before linear terms.

$4cd \times 3c^2 = 12c^3d$

$\frac{3x(x+y)}{6x^2} = \frac{x+y}{2x}$

When you divide (or multiply) the top and bottom of a fraction by the same number the result stays the same. In this case it was possible to divide top and bottom by 3x.

Simplifying by removing brackets

Make sure you understand how these have been done.

$4x - 2(2x - 5) = 4x - 4x + 10 = 10$

$c^2 - 3c(c - 4d) = c^2 - 3c^2 + 12cd = {}^-2c^2 + 12cd$

Expanding double brackets ► page 31

Simplifying by factorising

Factorising means writing an expression as a product.
This usually involves putting in brackets.

In the expression $^-2c^2 + 12cd$, $2c$ is a factor of both terms, so we can write $^-2c^2 + 12cd = 2c(^-c + 6d)$.

Factorising quadratics ► page 31

The rules of indices

Indices ► page 14

You need to know how to simplify an algebraic expression involving indices.
The rules for any numbers x, y, a and b are:

$(xy)^a = x^a y^a$ $\quad x^{-a} = \frac{1}{x^a}$ $\quad x^{\frac{1}{b}} = {}^b\sqrt{x}$ $\quad x^{\frac{a}{b}} = {}^b\sqrt{(x^a)}$

$x^a \times x^b = x^{a+b}$ $\quad x^a \div x^b = x^{a-b}$ $\quad (x^a)^b = x^{ab}$ $\quad x^0 = 1$

Check that you agree with these: $2x^2 \times 3x^5 = 6x^{(2+5)} = 6x^7$ $\quad 6x^5 \div 3x^3 = 2x^{5-3} = 2x^2$

$1 \div (5x^3)^2 = \frac{1}{25x^6} = \frac{1}{25}x^{-6}$ $\quad (x^4 \times y^{-2})^{\frac{1}{2}} = x^2y^{-1} = \frac{x^2}{y}$

Substituting numbers into an expression

Be particularly careful with negative numbers. For example, if $a = {}^-4$,

$a^2 - 3(a - 2) = (^-4)^2 - 3(^-4 - 2) = 16 - 3(^-6) = 16 + 18 = 34$

Negative numbers ► page 3

1 Simplify these algebraic expressions where possible.

(a) $ab + 2ba + a$ (b) $4x - 3y - 2(x + y)$ (c) $3a \times {}^-6a$ (d) $a^3 \div a^5$

(e) $5a^2 \times 2ab$ (f) $\frac{1}{2}cd^2 \times 8de^2$ (g) $x(y + 2) - y(2x - 3)$ (h) $(x^{\frac{5}{2}})^2$

(i) $\frac{2x^2y^2}{4xy}$ (j) $\frac{1}{3}x^2 \div 2x$ (k) $\frac{a^2b^4}{a^4b^2}$ (l) $2a \div \frac{4a^4}{a^3}$

(m) $^-c^2 - 3c(c + 7)$ (n) $3x^2 + 2x^3$ (o) $(2m^2)^3$ (p) $(3d)^2 \times d^3$

2 Factorise these completely.

(a) $6x + 12y$ (b) $3a + ab$ (c) $n^2 + 5n$ (d) $10x^2 + 5xy$

(e) $14x - 6x^2$ (f) $^-x^2 + 3x$ (g) $^-2x^3 - 10x^2$ (h) $\frac{1}{3}\pi r^2 h + \frac{4}{3}\pi r^3$

3 If $a = {}^-3, b = 5$ and $c = \frac{1}{2}$, work out the values of these expressions.

(a) $a(b + c)$ (b) $a^2 - b^2$ (c) $(3a)^3$ (d) $^-2b^2$

(e) $\frac{1}{2}(a - 2b)$ (f) $\frac{c - a}{b - a}$ (g) 3^a (h) $^-\frac{3}{5}(3a + 2b)$

(i) 25^c (j) $\left(\frac{b}{c}\right)^2 - 2a$ (k) b^a (l) $8^{\frac{1}{a}}$

4 Find the value of $\frac{ab}{c}$ when $a = 10^6, b = 10^{-3}$ and $c = 10^{-4}$.

5 The equation of a curve is $y = \frac{12}{3 - 2x}$. Calculate the value of y when

(a) $x = \frac{1}{2}$, (b) $x = {}^-3$.

MEG (SMP)

6 The formula for the time in seconds of one swing of a pendulum is $T = 2\pi\sqrt{\frac{l}{g}}$.
Calculate T when $l = 1{\cdot}5$ and $g = 9{\cdot}8$.

7 Police use the formula $s = \sqrt{(30fd)}$ to estimate the speed of a vehicle from skid marks where s is the speed in miles per hour, d is the length of the skid in feet and f is a number which depends on the weather and type of road.
Estimate the speed of a car which makes a skid mark 80 feet long. Use $f = 0{\cdot}4$.

8 The volume (V) of this solid is given by the formula

$$V = \frac{1}{8}ah(8b + \pi a).$$

Find the value of V when

$a = 4{\cdot}6$ cm,
$b = 9{\cdot}3$ cm,
$h = 12{\cdot}7$ cm.

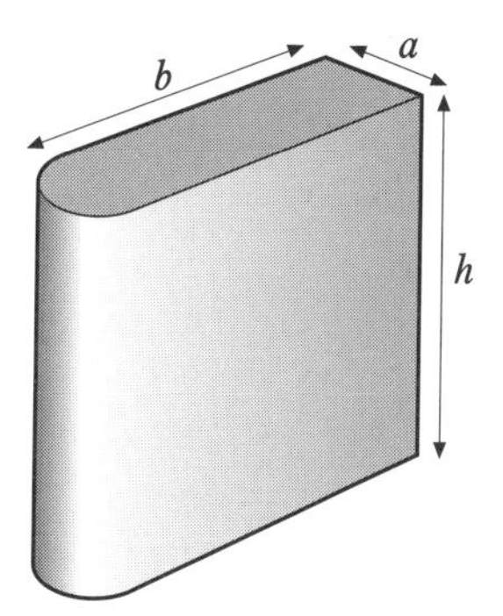

MEG/ULEAC (SMP)

9 (a) Factorise this expression completely: $2\pi rh + \pi r^2$

(b) A formula for the surface area of a cylinder with one end closed is $A = 2\pi rh + \pi r^2$.
Calculate the value of A when $h = 3\frac{3}{4}$ inches and $r = 1\frac{1}{2}$ inches.

MEG (SMP)

10 A formula used in statistics is

$$Z = \frac{X - m}{e} \quad \text{where} \quad e = \frac{s}{\sqrt{n}}$$

(a) Find the value of Z when $X = 14\frac{1}{2}, m = 16, s = 4\frac{1}{2}$ and $n = 25$.

(b) Find the value of Z when $X = 7{\cdot}81, m = {}^-3{\cdot}64, s = 14{\cdot}37$ and $n = 12$.

MEG/ULEAC (SMP)

Answers and hints ► page 107

Solving linear equations

Equations with x on one side of the equation only

To **solve** an equation we use the fact that it is still true if

- the same number is added to (or subtracted from) each side or
- each side is multiplied (or divided) by the same number.

Here are two ways to solve $\frac{x}{7} + 4 = 10$.

$\frac{x}{7} + 4 = 10$ *Subtract 4 from each side* ⇨ $\frac{x}{7} = 6$ *Multiply both sides by 7* ⇨ $x = 42$

$\frac{x}{7} + 4 = 10$ *Multiply both sides by 7* ⇨ $x + 28 = 70$ *Subtract 28 from each side* ⇨ $x = 42$

Check by substituting the value 42 for x in the left-hand side of the equation, that is $\frac{42}{7} + 4 = 6 + 4 = 10$, which equals the right-hand side.

Equations with x on both sides of the equation

Here is a way to solve an equation like $5x + 7 = 3x - 10$.

Subtract $3x$ from both sides. $2x + 7 = {}^-10$

Subtract 7 from both sides. $2x = {}^-17$

Divide both sides by 2. $x = \frac{{}^-17}{2} = {}^-8\frac{1}{2}$

Check that you understand this working.

Check by substituting in both $5x + 7$ and $3x - 10$: ${}^-42\frac{1}{2} + 7 = {}^-35\frac{1}{2}$ and ${}^-25\frac{1}{2} - 10 = {}^-35\frac{1}{2}$

1 Solve each of these equations.

(a) $3x - 2 = 10$ (b) $3a = 2$ (c) $\frac{x}{2} - 6 = {}^-1$ (d) $\frac{x-5}{3} = 3$

(e) $3w = 1{\cdot}8$ (f) $4x - 2 = 7 + 2x$ (g) $3x - 4 = {}^-10{\cdot}6$ (h) $\frac{1}{2}x + 5 = 2 + 3x$

(i) $\frac{2(2x-3)}{5} = {}^-2$ (j) $\frac{1}{3}(2x - 5) = 7 - x$ (k) $5(x - 3{\cdot}5) = 8$ (l) $5(x + 0{\cdot}4) = 2$

(m) $\frac{k}{0{\cdot}3} = 4{\cdot}9$ (n) $2z - \frac{3z}{5} = 14$ (o) $2(x - 9) = 3(3 - x)$ (p) $16 - 2(5 - 2x) = 1$

2 Ruchi is $\frac{1}{3}$ the age of her father. Ruchi is x years old.

(a) Write down, in terms of x,
(i) Father's age (ii) Ruchi's age 9 years ago (iii) Father's age 9 years ago

Nine years ago, Father was 6 times as old as Ruchi.

(b) (i) Using some of your answers to (a), write down an equation in x which represents the above statement.
(ii) Solve your equation to find the value of x.

ULEAC

3 Rectangle A has the same area as square B.
Find the length of the unknown side of rectangle A.

Answers and hints ► page 108

Looking at graphs

Straight line graphs

The measure of slope of a line is called its **gradient**.
It may be positive or negative, and can be calculated from the fraction $\frac{\text{vertical change}}{\text{horizontal change}}$.

For an equation in the form $y = mx + c$, the value of m gives the gradient and c gives the value of y where the line cuts the y-axis.

So if the line has a gradient of 2 and cuts the y-axis at the point (0, 3), its equation is $y = 2x + 3$.

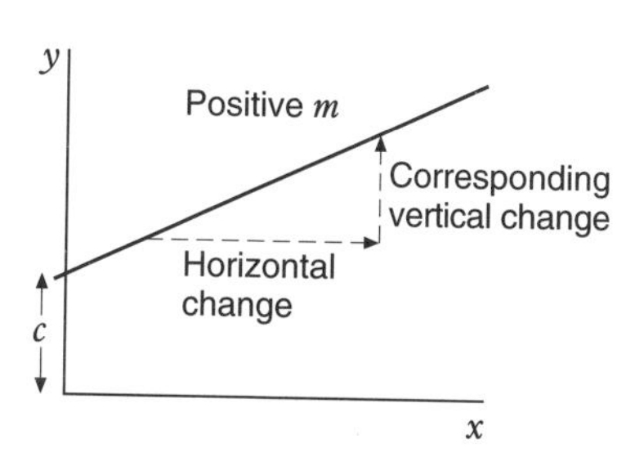

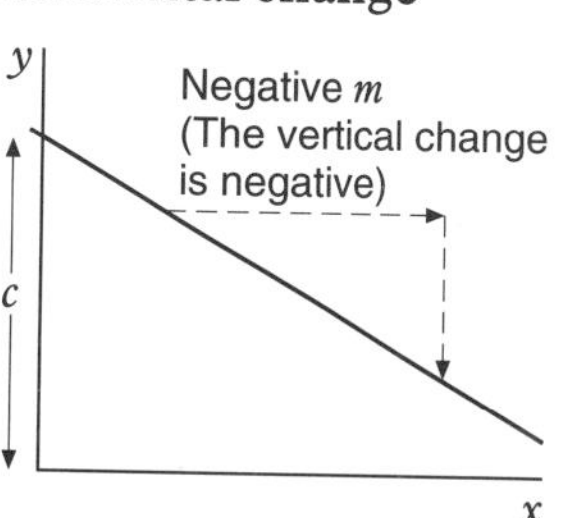

Non-linear graphs

Fitting functions to data ► page 40

You need to be able to recognise these graphs and sketch them from memory. (a and b are positive.)

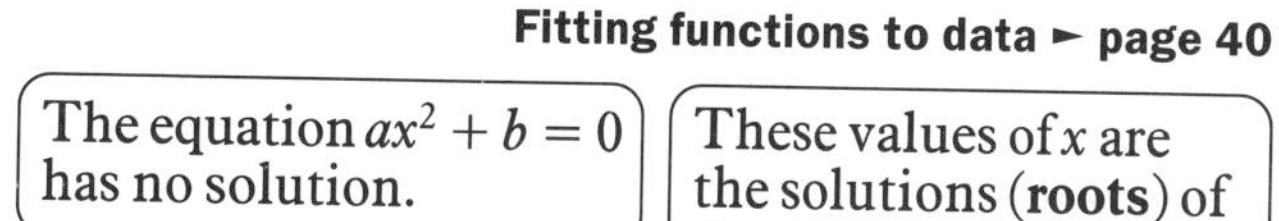

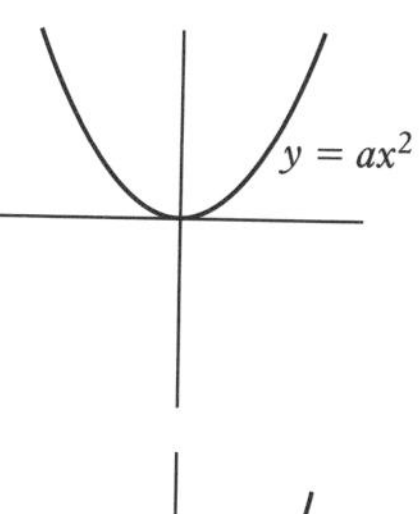

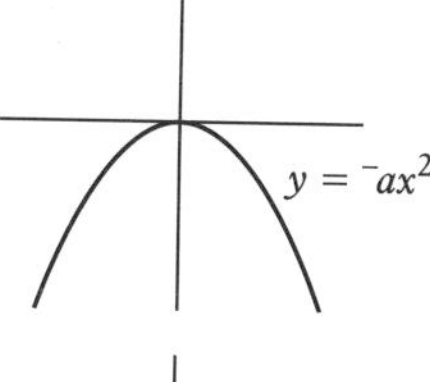

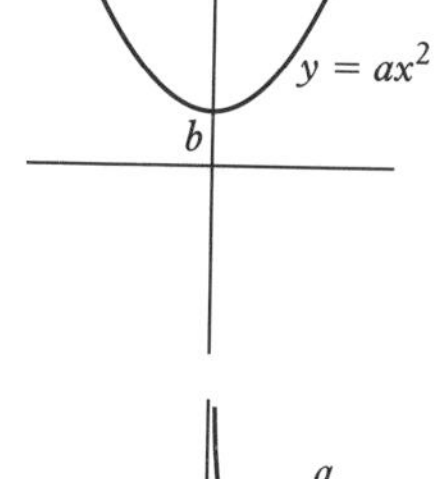

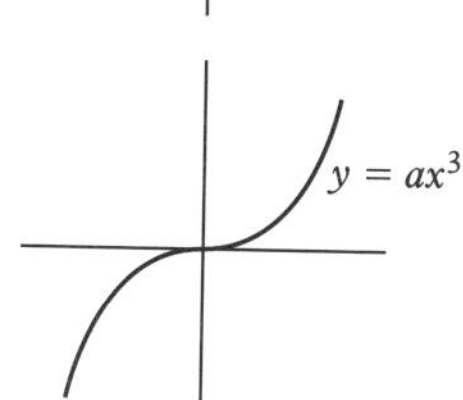

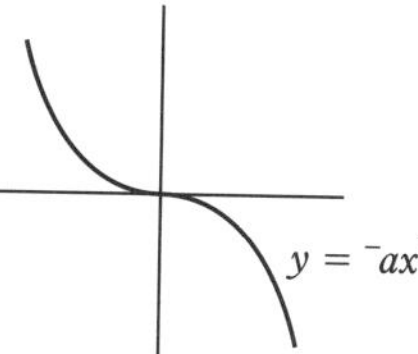

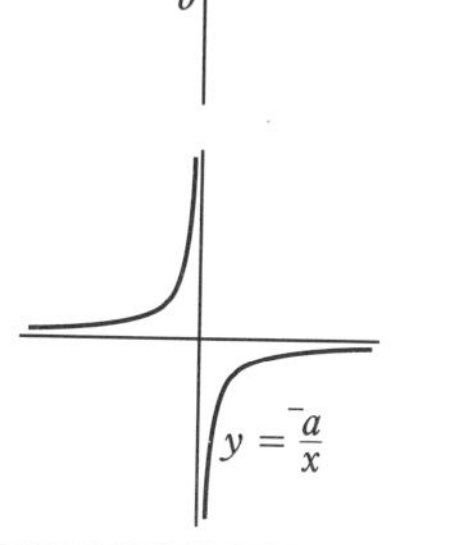

1 (a) The equation of a straight line is $2x + 3y = 12$.
Write this equation in the form $y = mx + c$.

(b) Write down the equation of the straight line through the origin which is parallel to the line $2x + 3y = 12$.

MEG (SMP)

2 Match the following equations to the graphs: $y = 2x^2 + 3$ $\quad y = 3 - 2x^2$ $\quad y = 2x^3$ $\quad y = \frac{2}{x}$

(a)

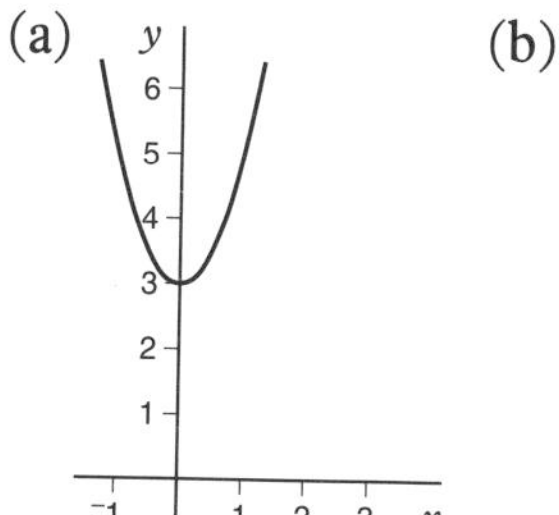

(b)

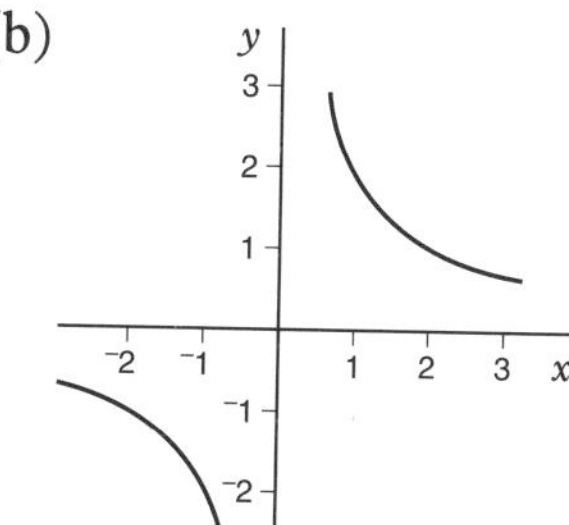

(c)

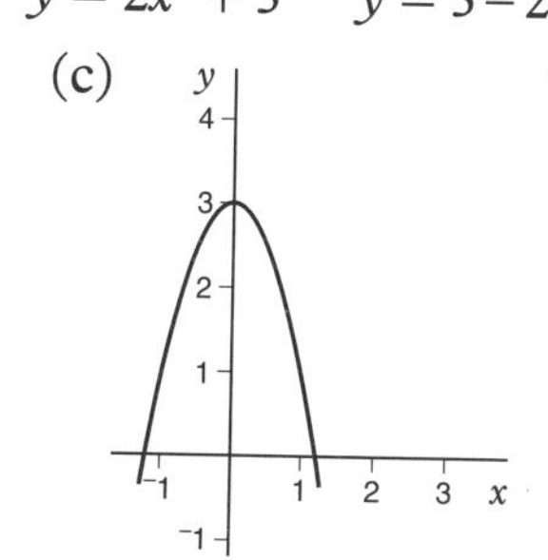

(d)

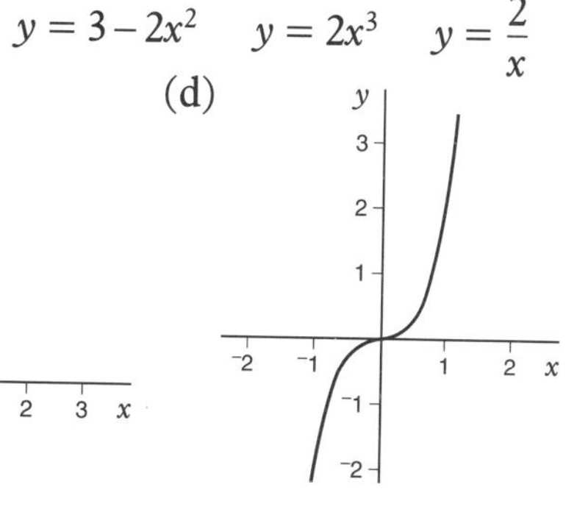

Answers and hints ► page 108

Simultaneous equations

Graphical solution

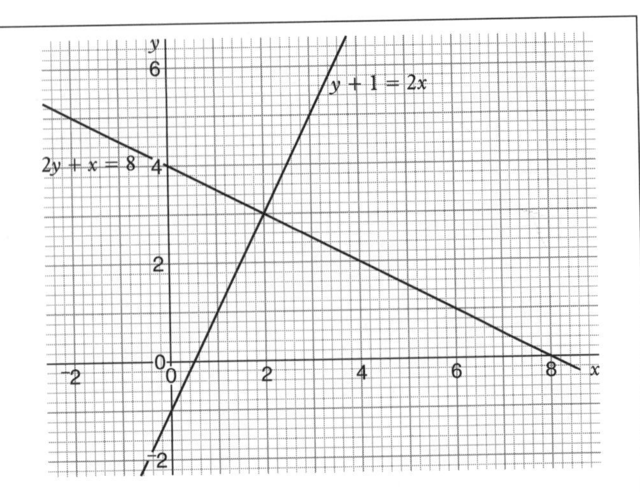

Look at the lines drawn on the grid.

There is only one point that fits both equations.

This is the point (2, 3) where the two lines cross.

$x = 2$ and $y = 3$ are the solutions to the pair of **simultaneous equations** $2y + x = 8$ and $y + 1 = 2x$.

Always check your solutions by substituting them back into the original equations.

Finding a solution without using a graph

Here are two methods of solving algebraically the simultaneous equations $2y + x = 8$ and $y + 1 = 2x$.

Method 1

Write out the two equations and label them.

$2y + x = 8$ *equation A*

$y + 1 = 2x$ *equation B*

If necessary rewrite one equation so that it is in the same form as the other.

$y - 2x = {}^{-}1$ *equation B rewritten*

⇕

If necessary multiply one equation to give a pair of equal terms in either x or y.

$4y + 2x = 16$ *equation A multiplied by 2*

Add or subtract both sides of the equations that you now have, to remove the equal terms.

$5y = 15$ *the two equations added together*

So $y = 3$

Substitute in equation A or B to find x or y.

$(2 \times 3) + x = 8$ *$y = 3$ substituted into equation A*

So $x = 2$

Always check your results by substituting them back in the 'other' equation:

Left-hand side of equation B $= 3 + 1 = 4$; right-hand side $= 2 \times 2 = 4$

Method 2

Here you substitute at the start. You need to be especially careful with the algebra.

From equation B we know that $y = 2x - 1$.

Substituting this expression for y in equation A gives

$$2(2x - 1) + x = 8$$
$$4x - 2 + x = 8$$
$$5x - 2 = 8$$
$$5x = 10$$
$$x = 2$$

Substitute $x = 2$ in equation B, say, to find the value of y: $y = (2 \times 2) - 1 = 3$

Check in A: left-hand side $= (2 \times 3) + 2 = 8$, which is the right-hand side.

You need 2 mm graph paper for questions 1 and 2.

1 Use a graphical method to solve the simultaneous equations

$y = 5 - x$
$y = 3x + 3$

(Draw axes with values of x and y from $^-5$ to 5 and a scale of 1 cm to 1 unit.)

2 Use a graphical method to solve the simultaneous equations

$2y = 3x + 4$ and $2x + 3y = 9$.

Give your answers to 1 decimal place.
(Draw axes with values of x and y from $^-2$ to 5 and a scale of 2 cm to 1 unit.) MEG/ULEAC (SMP)

3 Solve each of these pairs of simultaneous equations by the algebraic method of your choice. Check your answers.

(a) $3x + 4y = 33,\ x - 2y = 1$
(b) $x + 5y = 22,\ 3x + 2y = 14$
(c) $x - 3y = 9,\ 2x + y = 4$
(d) $2x + 7y = 47,\ 5x + 4y = 50$

4 At an indoor market in Blackpool there is a stall where they sell articles at either 50p or £1 each.
On 3 October they sold 229 articles and took £140.
Let the number of articles sold at 50p be x and the number of articles sold at £1 be y.

(a) Explain why $x + 2y = 280$.
(b) Write down another equation involving x and y.
(c) Solve the two equations algebraically and find how many articles were sold at each price. MEG/ULEAC (SMP)

5 I have chosen two numbers, a and b.
The sum of $2a$ and b is 34. The sum of a and $2b$ is 32.
Find the values of a and b.

6 The sum of the ages of a mother and her daughter is 43 years.
Four years ago, the mother was six times as old as her daughter.
What are their present ages?

7 (a) Solve the simultaneous equations

$3x + 2y = 12$
$4x - 3y = {}^-1$

(b) (i) Hence write down the value of p^{-1} and the value of $q^{\frac{1}{2}}$ if

$3p^{-1} + 2q^{\frac{1}{2}} = 12$
$4p^{-1} - 3q^{\frac{1}{2}} = {}^-1$

(ii) Find the value of p and the value of q. MEG

Answers and hints ► page 109

Inequalities and regions

A statement involving $>, \geq, <$ or $\leq$ is called an **inequality**.

Equations remain true if 'you do the same to both sides'.

The rule is not so simple for inequalities:

- the inequality remains true if we add the same number to both sides or subtract the same number;
- it also remains true if we multiply or divide by a positive number;
- however, if we **multiply or divide by a negative number, the direction of the inequality reverses.**

Solving linear equations ► page 22

Check that you can follow the solutions of these inequalities.

$$x > 8 + 2x$$

Subtract $2x$ from both sides. $\quad {}^{-}x > 8$

Multiply both sides by ${}^{-}1$, remembering to reverse the sign. $\quad x < {}^{-}8$

Check by substituting, say, $x = {}^{-}9$ into both sides of the original inequality.

$$2x^2 - 18 \leq 0$$

Add 18 to both sides. $\quad 2x^2 \leq 18$

Divide both sides by 2. $\quad x^2 \leq 9$

There are positive and negative solutions. $\quad {}^{-}3 \leq x \leq 3$

Check by substituting, say, $x = {}^{-}3$ and $x = 1$ into the original inequality.

Inequalities can be shown on a graph.
Check that the coordinates of all points above the line $y - \frac{1}{2}x = 1$ satisfy the inequality $y - \frac{1}{2}x > 1$ and points below the line satisfy $y - \frac{1}{2}x < 1$.

The line $y - \frac{1}{2}x = 1$ is the boundary between the two **regions** $y - \frac{1}{2}x > 1$ and $y - \frac{1}{2}x < 1$.

From the graph it is possible to see that points in the untinted region satisfy both the inequalities $y - \frac{1}{2}x < 1$ and $y > x^2 - 4x + 3$.

It is usually best to shade out the regions you do *not* want.

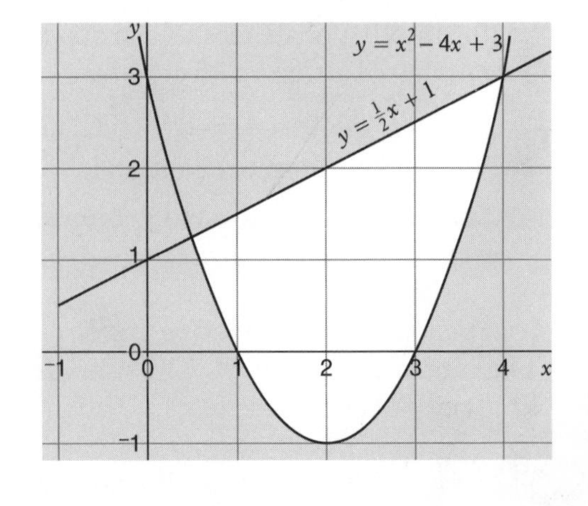

1 Solve these inequalities.

(a) $5x + 17 > 7$ (b) $x + 7 < 3x + 2$ (c) $3 - 2y < 11$ (d) $\frac{1}{2}(2 - 5n) \leq 11$

(e) $5 + 3x > \dfrac{x}{2}$ (f) $10 < 2 - 4x$ (g) $5(a - 2) - 8a \geq 0$ (h) $3x^2 - 12 > 0$

2 On graph paper, draw x- and y-axes, and mark each axis from ${}^{-}3$ to 5 using a scale of 2 centimetres to represent 1 unit.
On your diagram draw and label clearly the region which satisfies all of these inequalities.

$y \geq {}^{-}2$
$y \leq 3x + 1$
$2x + y \leq 5$

MEG

3 The values of x satisfy the inequality $3x + 1 \le 27 \le 5x - 6$.

(a) (i) Find the largest possible value of x.

(ii) Find the smallest possible value of x.

(b) Write down all the possible integer values of x.

MEG

4 (a) What can you say about x if $13 - 6x$ is less than 25?

(b) If $y - z \le 12$ and $z \le 7$, what is the greatest value of y?

5 The diagram shows the graphs of $y = x + 1$ and $y = x^2$.

(a) Write down two inequalities that describe the shaded area.

(b) At the point P, explain why $x^2 = x + 1$.

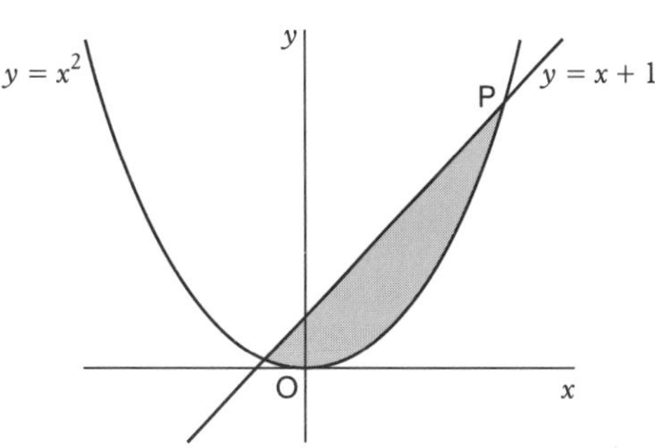

MEG (SMP)

6 Use this graph of $y = {}^{-}x^2 + 3x$ to help you decide which of these statements are correct.

(a) ${}^{-}x^2 + 3x \le 0$ for $0 < x < 3$

(b) ${}^{-}x^2 + 3x \ge 0$ for $0 \le x \le 3$

(c) ${}^{-}x^2 + 3x > 0$ for $0 \le x \le 3$

(d) ${}^{-}x^2 + 3x > 2$ for $1 \le x \le 2$

(e) ${}^{-}x^2 + 3x \ge 2$ for $1 \le x \le 2$

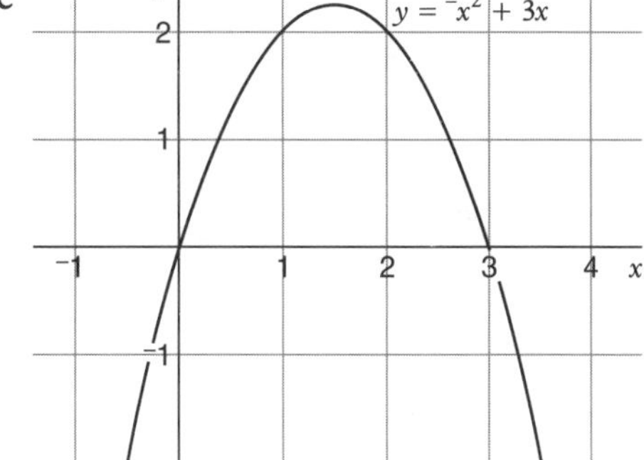

7 Nita has two types of fish in a tank: loach and guppy. There are x loach and y guppy.

(a) Nita has at least 10 guppy. Write this information as an inequality.

(b) The tank will accommodate up to 30 fish. Write this information as an inequality.

Draw a grid with values of x and y from 0 to 30 using a scale of 2 cm to 5 units.

(c) On the grid indicate the region which satisfies both these inequalities.

Nita has three times as many guppy as loach.

(d) Draw a line on the grid to show this information.

(e) How many of each sort of fish could Nita have? Circle the points on the grid which show all the possible answers.

MEG/ULEAC (SMP)

Answers and hints ► page 109

Trial and improvement; graphical solution

You need to be able to solve cubic equations like $x^3 - 5x = 2$.
Here are two methods you can use for solving cubic, and other, equations.

Using trial and improvement

In this method we try different values of x until one is found that gives $x^3 - 5x$ a value of 2.
Follow the working in the table, which starts with $x = 2$ as a reasonable first trial value.

You need to make this as close as possible to 2.

Substitution ► page 20
Significant figures ► page 4

Trial value of x	Value of $x^3 - 5x$
2	$^-2$
3	12
2·5	3·125
2·2	$^-0·352$
2·4	1·824
2·45	2·456

$x = 2$ is too small and $x = 3$ is too large, so the solution is between 2 and 3.

The solution is between 2·2 and 2·5. Try 2·4.

The solution is between 2·4 and 2·5. Try 2·45.

The solution is between 2·4 and 2·45. So the solution to 2 s.f. is 2·4.

Check that the solution correct to 3 s.f. is 2·41.

If you are careful this should take only three more trials.

Using graphs

An alternative method for solving $x^3 - 5x = 2$ is to use graph paper or a graphical calculator.

- Plot the graphs $y = x^3 - 5x$ and $y = 2$ on the same grid.
- Find the x-values of the points where the two graphs intersect. These are the solutions to the equation $x^3 - 5x = 2$.

As you can see, the equation has three solutions.

Drawing a graph is a good way to get the three approximate solutions which you could then find more accurately by trial and improvement.

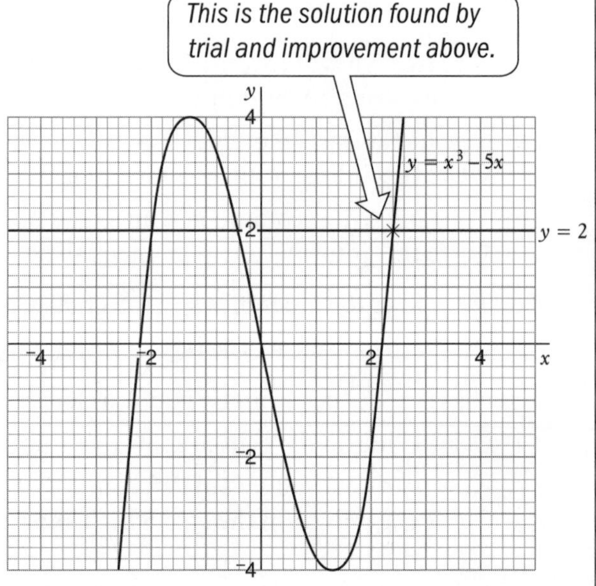

1 The equation $x^3 - 5x^2 + x - 10 = 0$ has a solution between 5 and 6.
Use trial and improvement to find this solution correct to 3 significant figures.
You must show all your trials.

2 Use trial and improvement to solve the equation $2x^3 = 15$ correct to 2 decimal places.

3 (a) Copy and complete this table of values for $y = x^3 - 3x - 2$.

x	$^-3$	$^-2$	$^-1$	0	1	2	3
y	$^-20$		0		$^-4$		

Draw axes with x-values from $^-3$ to 3 (2 cm to 1 unit)
and y-values from $^-20$ to 20 (2 cm to 5 units).

(b) Draw the graph of $y = x^3 - 3x - 2$.

(c) By drawing a suitable straight line, solve the equation $x^3 - 3x - 2 = 3$.

(d) Use trial and improvement to solve the equation
$x^3 - 3x - 2 = 25$ correct to 1 decimal place.

4 (a) Draw the graphs with equations $y = x^2$ and $y = 3x + 2$ for values of x between $^-4$ and 4.

(b) Use the graphs to solve the equation $x^2 = 3x + 2$.

(c) (i) On your grid draw the graph necessary to find the positive solution to the equation
$\frac{2}{x} = 3x + 2$.

(ii) Write down the positive solution to the equation $\frac{2}{x} = 3x + 2$.

(d) Find the equation of the straight line graph that it is necessary to draw
to solve the equation $x^2 + 4x - 1 = 0$.
(You do not need to draw the graph.)

MEG

5 (a) Given that $y = \frac{x^2}{10} - \frac{1}{x}$,
copy and complete the table.

x	0·5	1	2	3	4	5	6
$\frac{x^2}{10}$	0·025						3·6
$\frac{^-1}{x}$	$^-2$						$^-0·1\dot{6}$
y	$^-1·975$						$3·4\dot{3}$

(b) Draw the graph of $y = \frac{x^2}{10} - \frac{1}{x}$
for values of x from 0·5 to 6.
Use x-values from 0 to 6
(2 cm to 1 unit) and y-values
from $^-2$ to 4 (4 cm to 1 unit).

(c) (i) Explain clearly how you can find an estimate of $\sqrt[3]{10}$ from the graph.
(ii) Using your graph, find an estimate of $\sqrt[3]{10}$.

(d) Using the same axes and scales, draw the graph of $y = 3 - 2x$.

(e) Write down the value of x where the two graphs intersect.

(f) Show that you can use these two graphs to find an approximate solution
of the equation $x^3 + 20x^2 - 30x - 10 = 0$.

ULEAC

Answers and hints ► page 110

Gradients and tangents

A man cycles up a hill, rests at the top and then 'freewheels' down the other side.
The graph shows the distance he cycled during these three stages of his journey.

He cycled up the hill at a constant **rate**.
The graph for this section is a sloping straight line.
It shows that he travelled 1200 m in 5 minutes.
So his speed was $1200 \div 300 = 4$ m/s.

The cyclist rested for 4 minutes.
This part of the graph is a straight horizontal line – his speed is 0 m/s!

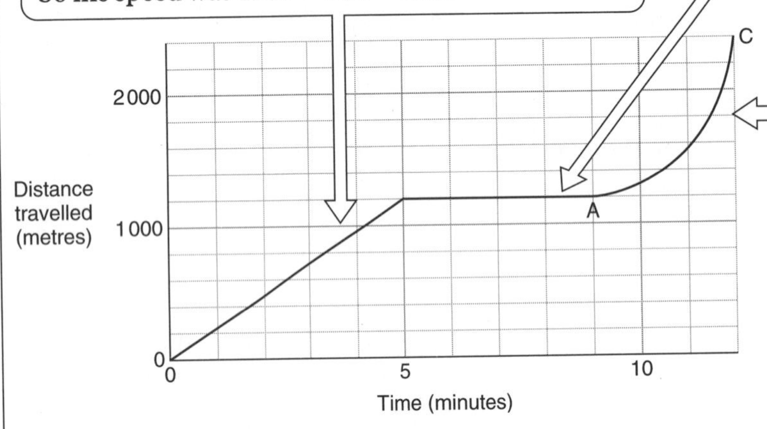

For the last part, down the hill, he was getting faster and faster, not cycling at a constant speed.
The fact that the graph is curved shows that his speed was changing.

Straight line graphs ► page 23

This is the distance–time graph for the last part of the journey.
The *average* speed in m/s between A and C is

$$\frac{2400-1200}{(12-9)\times 60} = \frac{1200}{180} = 6\tfrac{2}{3}.$$

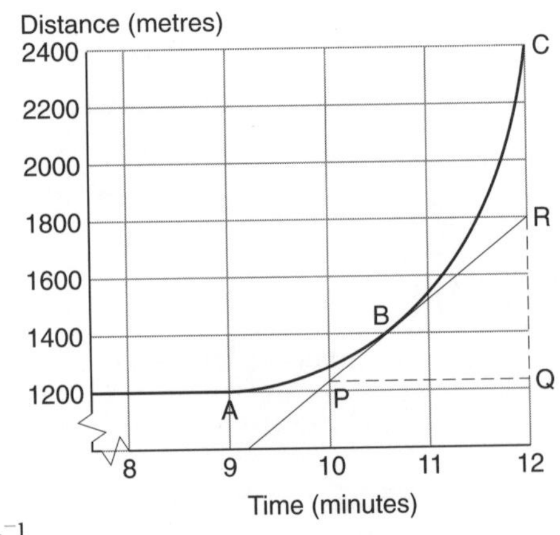

To find the cyclist's speed after, say 10·5 minutes, draw the **tangent** at B (a line which *touches* the curve at B).
The gradient of this line gives the speed at B.

Calculate the gradient by drawing a suitable triangle.
(For accuracy use as big a triangle as reasonable with a base a convenient number of units long.)
Here the triangle gives a speed of $\frac{570}{120} \approx 4{\cdot}75$ m/s or $4{\cdot}75\,\text{m s}^{-1}$.

The rate of change of speed with time gives the **acceleration**.
For the cyclist above this would be measured in metres per second/second which is written as m/s^2 or m s^{-2}.

This graph shows how a car's speed varies as it draws away from rest. To find the acceleration after, say 5 seconds, draw a tangent at P and then a suitable triangle ABC.
Measure the increase in speed (about 50 km/h) and the time that has elapsed (10 seconds), and so find

the acceleration $\approx \dfrac{50\,\text{km/h}}{10\,\text{sec}} = \dfrac{(50\times 1000)\,\text{m}}{(60\times 60)\,\text{s}} \times \dfrac{1}{10\,\text{s}} \approx 1{\cdot}4\,\text{m/s}^2.$

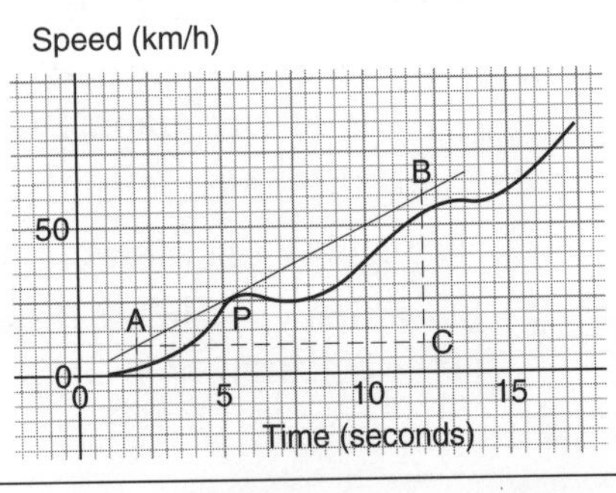

All the questions are on worksheets H1 and H2.

Answers and hints ► page 112

Quadratics

Expanding double brackets

A multiplication grid is useful for multiplying out an expression like $(x + 2)(x - 5)$. It means there is less risk of missing out terms.

$\times$	x	$^-5$
x	x^2	^-5x
2	$2x$	$^-10$

Total $= x^2 - 5x + 2x - 10$
$= x^2 - 3x - 10$

So $(x + 2)(x - 5) = x^2 - 3x - 10$.

Solving quadratic equations by factors

Quadratic equations like $x^2 + 5x - 6 = 0$, $z^2 + 3z = 0$ and $x^2 + 6 = 5x$ can have two different solutions or **roots**.

Graphical solution of equations ► page 28

One way to solve a quadratic is by **factorising**.

Example Solve $4x^2 - 12x = 0$.

Factorising gives $4x(x - 3) = 0$.

If the product of two numbers is zero, then one or both of them must be zero.

So either $4x = 0$ or $x - 3 = 0$.
This means that $x = 0$ and $x = 3$ are both solutions of $4x^2 - 12x = 0$.
(Check by substituting these values back in the equation.)

Example Solve $x^2 + 5x - 6 = 0$ by factorisation.

*As the constant term is **subtracted**, look for two numbers whose product is 6 and whose **difference** is 5.*

Factorising gives $(x - 1)(x + 6) = 0$.
So either $x - 1 = 0$ or $x + 6 = 0$.
The solutions are $x = 1$ and $x = {}^-6$. (Check by substituting back.)

Solving quadratic equations by using the standard formula

Any quadratic equation $ax^2 + bx + c = 0$ can be solved by using the standard formula

$$x = \frac{^-b \pm \sqrt{(b^2 - 4ac)}}{2a}.$$

Remember to divide the whole of the numerator by 2a.

Example Find the roots of the equation $2x^2 - x - 5 = 0$.

Substitute $a = 2, b = {}^-1$ and $c = {}^-5$ into the standard formula.

$$x = \frac{^-(^-1) \pm \sqrt{(1 + 40)}}{4} = \frac{1 \pm \sqrt{41}}{4} = 1{\cdot}85 \text{ or } {}^-1{\cdot}35 \text{ (to 2 d.p.)}$$

1 Multiply out these brackets and simplify each expression.

(a) $(2x - 1)(x + 7)$ (b) $(3n - 1)(3n + 1)$ (c) $(3x - 2)^2$ (d) $(5a + 7)(7 - 4a)$

2 Solve these quadratic equations by factorising.

(a) $x^2 + 4x = 0$ (b) $x^2 + 7x + 12 = 0$ (c) $x^2 - 5x - 14 = 0$
(d) $x^2 - x = 12$ (e) $x^2 + 2x - 80 = 0$ (f) $x^2 + 6 = 3x + 60$
(g) $x^2 = 28 - 3x$ (h) $3(2x^2 + x) = 12 - 3x$ (i) $(x + 3)^2 = 2(x + 6) + 9$

3 Use the formula to solve these these equations correct to 2 decimal places.

(a) $x^2 - 15x + 50 = 0$ (b) $x^2 + 20x - 32 = 0$ (c) $x^2 + 15x - 50 = 0$
(d) $2x^2 + 12x - 85 = 0$ (e) $2x^2 - 5x = 2$ (f) $3x^2 = 8x + 8$

Answers and hints ► page 112

Algebraic fractions

The main technique used to simplify expressions that include fractions is to multiply both numerator and denominator by the same expression so that all fractions have the same denominator.

Example Simplify $\frac{x}{3y} - \frac{y}{4x}$. ⇐ *Both fractions must have the same denominator.*

Multiply the first fraction by $\frac{4x}{4x}$ and the second by $\frac{3y}{3y}$ so that both fractions have the denominator $12xy$.

The expression becomes $\frac{4x^2}{12xy} - \frac{3y^2}{12xy}$ which can be rewritten as $\frac{4x^2 - 3y^2}{12xy}$.

Example Write as a single algebraic fraction $\frac{3x+1}{x-2} - \frac{x}{2x+3}$.

Making $(x-2)(2x+3)$ the denominator for both fractions means you can write the expression as

$$\frac{(3x+1)(2x+3) - x(x-2)}{(x-2)(2x+3)} = \frac{(6x^2+11x+3) - (x^2-2x)}{(x-2)(2x+3)}$$

⇐ *The first fraction is multiplied by* $\frac{2x+3}{2x+3}$ *and the second by* $\frac{x-2}{x-2}$.

When the brackets in the numerator are removed this becomes

$$\frac{6x^2+11x+3-x^2+2x}{(x-2)(2x+3)} \quad \text{which is} \quad \frac{5x^2+13x+3}{(x-2)(2x+3)}.$$

⇐ *$5x^2 + 13x + 3$ does not factorise.*

Example Solve $\frac{3}{x-1} + \frac{4x-3}{2x+3} = 2$.

$$\frac{3(2x+3) + (4x-3)(x-1)}{(x-1)(2x+3)} = 2.$$

$$\frac{4x^2 - x + 12}{2x^2 + x - 3} = 2$$

⇐ *Remove the brackets and collect terms.*

Now rearrange the equation on one line by multiplying both sides by $2x^2 + x - 3$:

$4x^2 - x + 12 = 4x^2 + 2x - 6$ which simplifies to $0 = 3x - 18x$

Solving this equation gives $x = 6$

1 Write these as single algebraic fractions (leaving the denominator as a product of factors).

(a) $\frac{1}{x} + \frac{1}{y}$ (b) $\frac{2}{x+4} + \frac{1}{x+1}$ (c) $\frac{3}{x-4} - \frac{2}{x+3}$ (d) $\frac{4x}{x-2} - \frac{x-5}{x+7}$ (e) $\frac{3x}{y} + \frac{y}{x}$

2 Solve these equations.

(a) $\frac{3}{1-2z} + \frac{9}{10} = \frac{3}{2}$ (b) $\frac{2}{x-3} + \frac{5}{x+3} = 1$ (c) $\frac{x}{x+1} + \frac{2x}{x-2} = 3$ (d) $\frac{y+1}{y-1} - \frac{2y}{y+8} = 1$

3 (a) Show that the equation $\frac{2}{x-1} - \frac{3}{x+2} = 4$ can be written in the form $4x^2 + 5x - 15 = 0$.

(b) Solve this equation giving your answers correct to 2 decimal places.

Answers and hints ► page 113

Changing the subject of a formula

A **formula** can be seen as an equation with more than one variable (letter).

In the formula $t = \frac{2p + r}{p - 2r}$, t is the **subject** of the formula.

Rewriting so that a different variable is the subject is just like *solving* for that variable.

Example Make b the subject of the formula $a = \frac{7 + \sqrt{b}}{c}$.

Multiply both sides by c. $ac = 7 + \sqrt{b}$
Rearrange so that $\sqrt{b}$ is on one side. $\sqrt{b} = ac - 7$
Square both sides. $b = (ac - 7)^2$

Example Make r the subject of the formula $t = \frac{2p + r}{p - 2r}$.

Multiply both sides by $p - 2r$. $t(p - 2r) = 2p + r$
Remove the brackets. $tp - 2tr = 2p + r$
Rearrange so that r is on one side only. $tp - 2p = r + 2tr$
Factorise. $p(t - 2) = r(1 + 2t)$
Divide both sides by $(1 + 2t)$. $r = \frac{p(t-2)}{1 + 2t}$

Algebraic fractions ► page 32

1 Rearrange each of these formulas to make the letter in square brackets the subject.

(a) $\frac{K}{T} = 3V^2$ [T] (b) $b = 2k\sqrt{a}$ [a] (c) $v^2 = u^2 + 2as$ [a] (d) $s = \frac{v^2 - u^2}{2a}$ [v]

(e) $m = \frac{1}{p}\sqrt{(n^2 - p^2)}$ [p] (f) $xt^2 - \frac{k(t+1)}{x} = 0$ [x] (g) $k = \frac{rh}{r + h}$ [h] (h) $\frac{1}{u} + \frac{1}{v} = \frac{1}{f}$ [u]

2 A body of mass m kilograms, moving with an initial velocity of u metres per second, is acted on by a force of F newtons for a time t seconds.
At the end of this time, its velocity has changed to v metres per second.
The formula connecting these variables is $mv - mu = Ft$.

(a) Make m the subject of the formula.

(b) Find m when $v = 7{\cdot}23$, $u = {}^{-}3{\cdot}17$, $F = 52{\cdot}00$ and $t = 0{\cdot}76$.

3 $y = \frac{4}{7 - x} + 6$

(a) Find the value of y when $x = 3\frac{3}{4}$. (b) Find the value of x when $y = {}^{-}3$. MEG/ULEAC (SMP)

4 A formula about the path of atoms in a magnetic field is $r = \frac{Mv}{2eH}$.

(a) Find the value of r when $M = 2{\cdot}01 \times 10^{-23}$, $v = 2{\cdot}0 \times 10^7$, $e = 1{\cdot}6 \times 10^{-20}$ and $H = 2000$.

(b) Rearrange the formula to make v the subject. MEG/ULEAC (SMP)

5 An equation for x in the form $x^3 + px^2 + qx + r = 0$ can be derived from $x = \sqrt{\left(3 + \frac{1}{x}\right)}$.

Find the values of p, q and r. MEG/ULEAC (SMP)

Answers and hints ► page 114

Using algebra

Sequences and terms

Suppose we are told that the nth term of a particular sequence is $n^2 - 1$.
This means that the 1st term is $1^2 - 1 = 0$,
the 2nd term is $2^2 - 1 = 3$,
the 3rd term is $3^2 - 1 = 8$, and so on.
So the sequence is 0, 3, 8, 15, 24, ...

It helps to look for factors, square and cube numbers, numbers going up by the same amount, ... when you are looking for patterns in sequences.

Convince yourself that for the sequence
1, 3, 7, 15, 31, ... the nth term is $2^n - 1$.

Solving problems

Solving linear equations ► page 22

Solving quadratic equations by factors ► page 31

Solving simultaneous equations without a graph ► page 24

Nick is thinking of a number. If he multiplies it by 8 and subtracts 5 he gets the same as the square of the number plus 10.
What number is Nick thinking of?

Let the number be x. We need to form an equation in x and solve it.

$8x - 5 = x^2 + 10$
$x^2 - 8x + 15 = 0$
$(x - 3)(x - 5) = 0$
$x = 3$ or 5.

So Nick must be thinking of the number 3 or 5.

Check by substituting in the original explanation, not your equation in case it is wrong.

1 Write the first five terms of the sequence whose nth term is

(a) $4n - 1$ (b) $n^2 + 3$ (c) $2^{n-1} + 1$ (d) $n(n - 1)$

2 Find in terms of n, the nth term of each of these sequences.

(a) 3, 6, 9, 12, 15, ... (b) 3, 9, 27, 81, ...
(c) 3, 6, 12, 24, 48, ... (d) 2, 8, 18, 32, 50, ...

3 Write down an algebraic expression for the results of multiplying the square of x by 5 and adding 3 times x.

4 Ravinder is thinking of a positive number.
If she doubles it and adds 18 it is equal to the square of the difference between the number and three.
What number is Ravinder thinking of?

5 Maria has two brothers, Aaron and Jason.
Maria is x years old.
Aaron is three years older than Maria.
Jason is two years younger than Maria.

(a) Given that the product of the two brothers' ages is 126, show that $x^2 + x - 132 = 0$.

(b) Use the equation to find the ages of the three children.

MEG (SMP)

6 Glyn is making rectangular patterns from counters.
The length of each rectangle is one more than the width.
The first three patterns are shown.

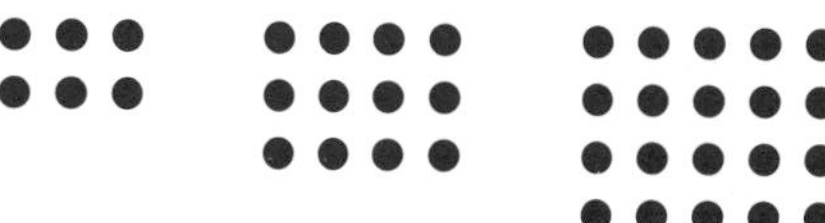

Each time, to get the next pattern, he makes the width one more counter.

(a) Write an expression for the number of counters in the nth pattern.

Glyn notices that there is also a pattern in how many extra counters he needs each time to make the next pattern.

(b) How many extra counters does he need to make the $(n + 1)$th pattern from the nth pattern?

MEG/ULEAC (SMP)

7 James wants to make a screen round part of his garden against a straight fence. He buys a 30 m roll of netting and uses it to make three sides of a rectangle.
(The other side is the fence.)
The length of one side is x m, as shown in the diagram.

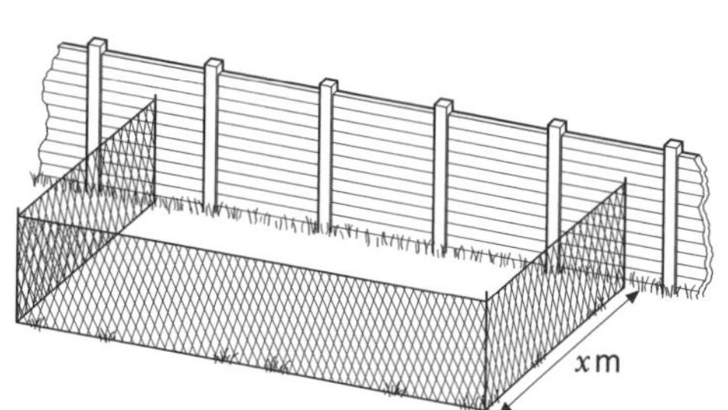

(a) Write down an expression for the area of garden inside the screen in square metres.

The area inside the screen is $100\,\text{m}^2$.

(b) Show that $x^2 - 15x + 50 = 0$ and solve the equation.

(c) Comment on your solutions.

MEG/ULEAC (SMP)

8 If n is a positive integer, then $n - 1$, n and $n + 1$ are three consecutive integers.
The sum of the squares of the integers is 365.
Form an equation in n and solve it to find the three integers.

9 The sum of the first n terms of the sequence 4, 10, 16, 22, ... is $n(3n + 1)$.
How many terms add up to 1344?

10 This rectangle has a perimeter of 16 cm and an area of $10\,\text{cm}^2$.
The length of one side is a cm.

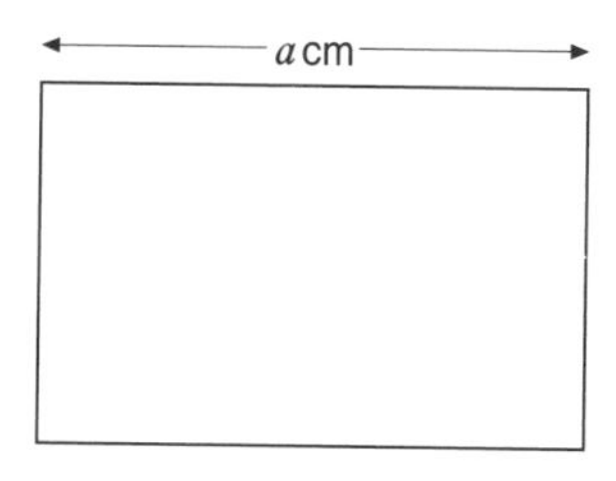

(a) Show how this information can be used to form the equation $a^2 - 8a + 10 = 0$.

(b) Solve the equation to find the length and width of the rectangle.
Give your answers to an appropriate degree of accuracy.

11 12, 13, 14 and 15 are four consecutive numbers.
If you multiply the outer pair you get 180.
If you multiply the inner pair you get 182.

(a) Choose three more sets of four consecutive numbers and carry out similar calculations.
What do you notice?

(b) Use algebra to prove that your rule always works with any four consecutive numbers.

(c) Use algebra to find a similar rule if you start with four consecutive *odd* numbers.

Answers and hints ► page 114

Functions and graphs

You can think of a **function** as an instruction which changes an **input** to give an **output**.
For example a function, f, might stand for 'multiply by 2 and add 3'.

So $f(4) = 2 \times 4 + 3 = 11$, $f(x) = 2x + 3$ and $f(x-1) = 2(x-1) + 3 = 2x + 1$.

The symbol f(4) ('f of four') means the output when 4 has been put in.

Translation parallel to the x-axis

Look at the graph below. The solid line is the graph of $f(x) = x^2 - 3x - 4$.
We want to find the function, g, which is shown by the dashed line.
This is the graph of $y = f(x)$ translated 3 units to the **right** parallel to the x-axis.

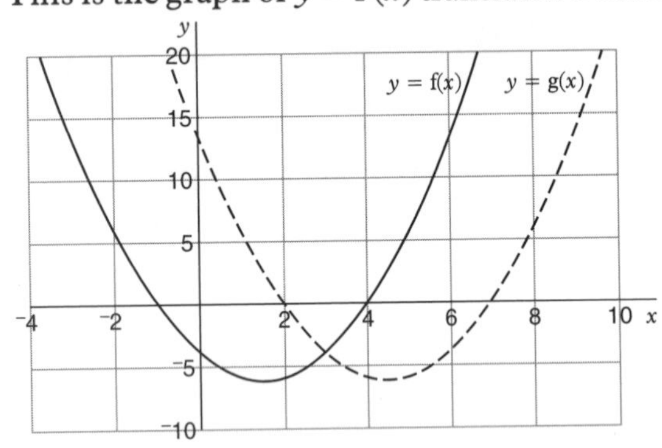

The diagram shows that $g(2) = f(^-1)$, that $g(7) = f(4)$, that $g(4\frac{1}{2}) = f(1\frac{1}{2})$, and so on.
For all values of x, $g(x) = f(x-3)$.
This can be rewritten in terms of x.

$$\begin{aligned} f(x-3) &= (x-3)^2 - 3(x-3) - 4 \\ &= (x^2 - 6x + 9) - (3x - 9) - 4 \\ &= x^2 - 9x + 14 \end{aligned}$$

So $g(x) = x^2 - 9x + 14$

Check that g(2) = 0 and g(7) = 0.

Check that for a translation 2 units to the **left**, the new function $h(x) = f(x+2) = x^2 + x - 6$.

Translation parallel to the y-axis

These are the graphs of $f(x) = \frac{1}{2}x^3 - 5x - 1$ and $g(x) = \frac{1}{2}x^3 - 5x + 3$.

The translation has vector $\begin{bmatrix} 0 \\ 4 \end{bmatrix}$ and $g(x) = f(x) + 4$.

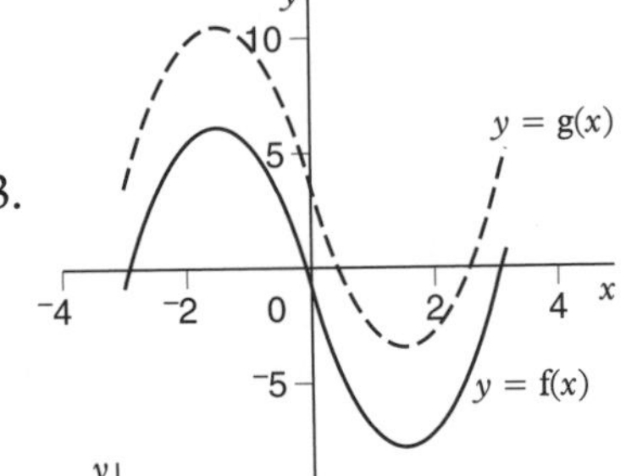

Stretches

If as in this diagram $g(x) = f(4x)$, then the graph of f is 'squeezed' to $\frac{1}{4}$ of its width parallel to the x-axis.
Note that $f(x)$ and $g(x)$ have the same value at $x = 0$.
If $h(x) = 4f(x)$, the graph is 'stretched' by a factor of 4 parallel to the y-axis.
$f(x)$ and $h(x)$ have the same value at $y = 0$.

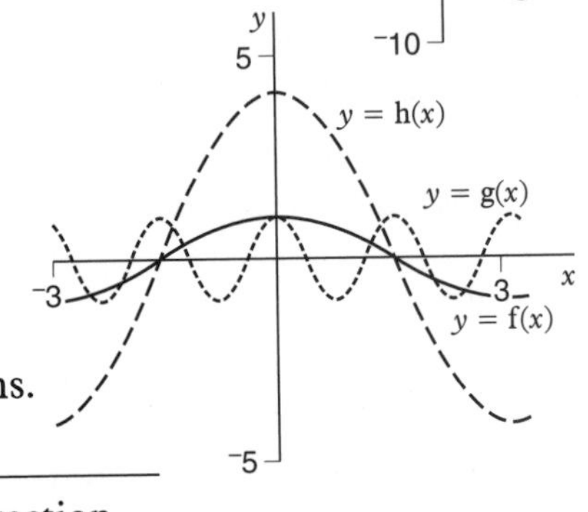

The table below summarises the results in general terms.

Function	Transformation of the graph of $y = f(x)$
$f(x-k)$	translation k units in the positive x-axis direction
$f(x) + k$	translation k units in the positive y-axis direction
$f\left(\frac{x}{k}\right)$	stretch in the positive x-axis direction, scale factor k
$f(kx)$	stretch in the positive y-axis direction, scale factor k

1 For the function $f(x) = (x-1)^2 + 4$, work out

(a) $f(5)$ (b) $f(1)$ (c) $f(a+1)$ (d) $f(^-3)$ (e) $f(\frac{1}{2})$

2 If $f(x) = 2x + 3$ and $g(x) = 4 - 3x$, find the value of x for which

(a) $f(x+1) = g(x-1)$ (b) $f(2x) = 3g(x)$ (c) $\frac{1}{2}f(x) = g(x+5)$

3 Draw the graph of the function $f(x) = 2x^2 + 3x - 9$.
The function g has the same graph but translated 2 units to the right.
The function h has the same graph but translated 1 unit to the left.
Sketch the graphs of $y = g(x)$ and $y = h(x)$.
Work out the equations for each of these functions, and check from these that the graphs cross the x-axis at the expected points.

4 The solid line is the graph of $f(x) = {}^-x^2 + 8x - 12$ and the dashed line is of $g(x) = {}^-x^2 - 6x - 5$.
Show that each graph is an exact translation of the other.

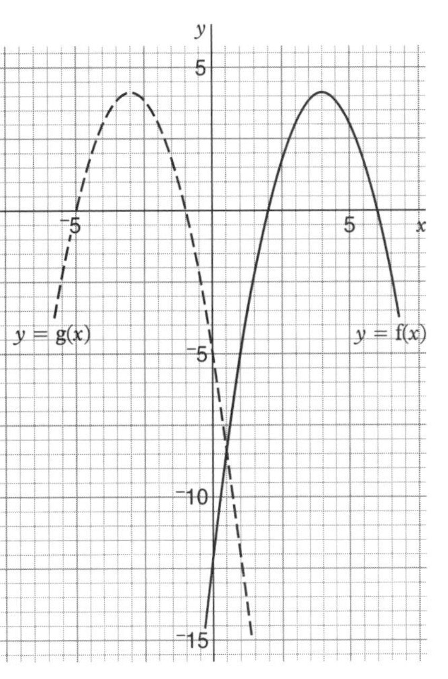

5 The three graphs shown in broken lines are each translations of the original $y = f(x)$.
If each of them is written in the form $y = f(x+a) + b$, what are the values of a and b each time?

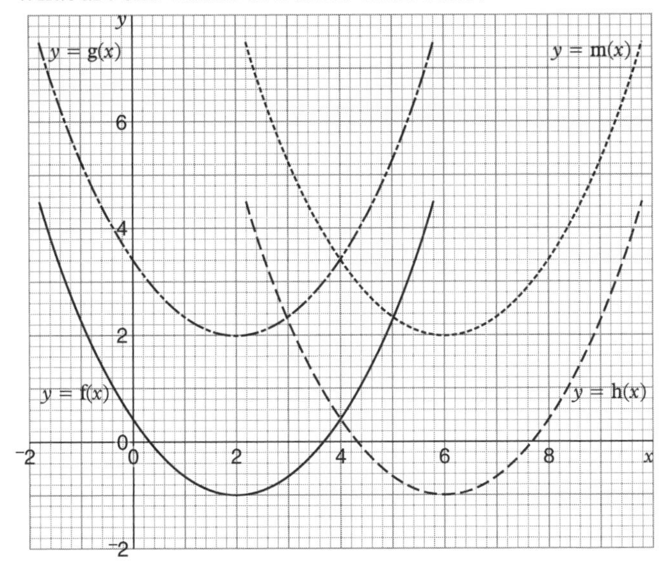

6 This is the graph of a function $y = f(x)$.
Sketch the graphs of the following.

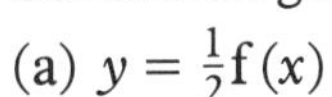

(a) $y = \frac{1}{2}f(x)$

(b) $y = 1 + f(x)$

(c) $y = f\left(\frac{x}{2}\right)$

(d) $y = f(x-1)$

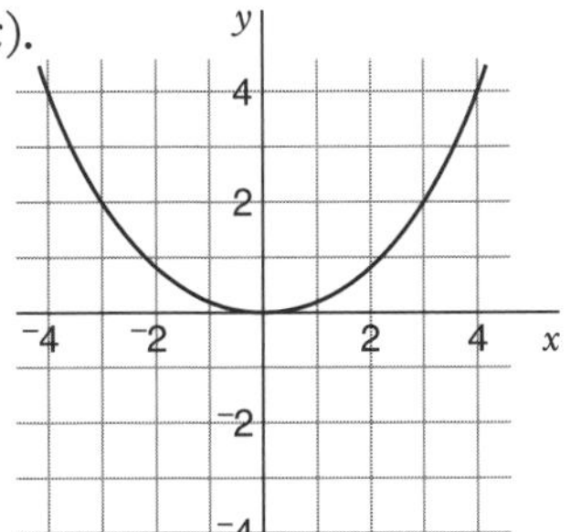

MEG/ULEAC (SMP)

Answers and hints ► page 116

The area under a graph

The formula for calculating the energy consumed by a piece of electrical equipment is: energy = power × time.

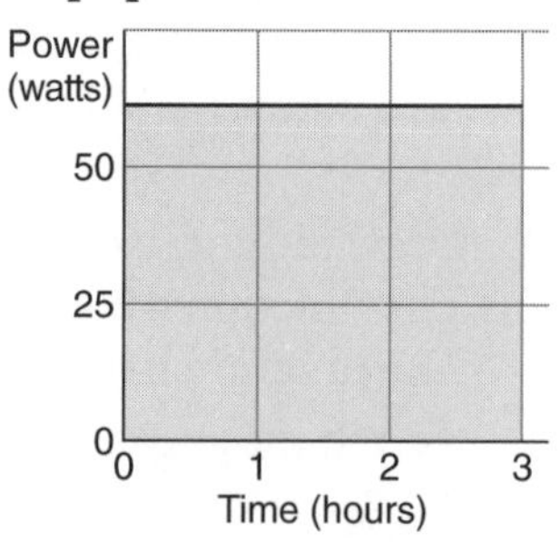

A light bulb of power 60 watts will consume 180 units (watt-hours) of energy in 3 hours. The graph shows the power consumption and the shaded area represents the energy used.

If the vertical axis was labelled 'speed' and not 'power' the graph would represent a car travelling at 60 m.p.h. for 3 hours; the shaded area would show how far it had travelled.

The same ideas hold for areas under curved lines.
This graph shows a vehicle moving off from standing still.
The shaded area shows how far it has travelled in the first 15 seconds.

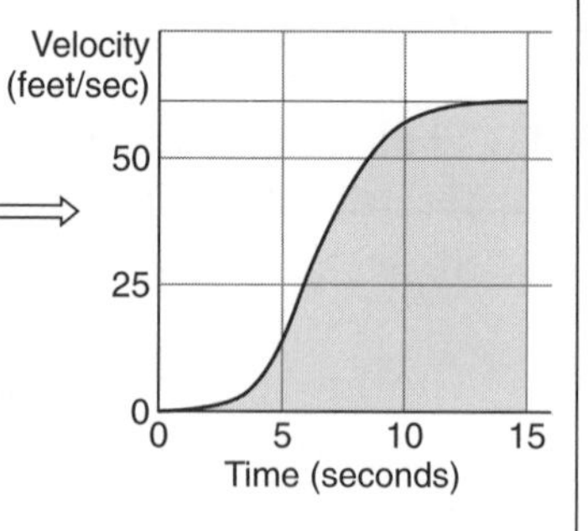

This is an enlarged view of the area under the graph between 3 and 4 seconds.
The area between the curve (DC) and the horizontal AB is more than the area of the rectangle ABPD, and less than the area of ABCQ.
It is close to the area of the trapezium ABCD:
$\frac{1}{2}$ × width of AB × (height of AD + height of BC)

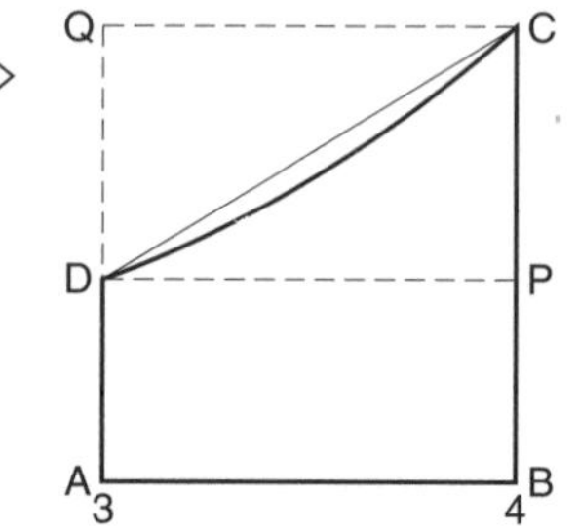

The **trapezium rule** is used for calculating the approximate area under a curve.
The area is divided into trapeziums of equal width w, and is given by
$\frac{1}{2} \times w \times$ (first height + 2 × each intermediate height + last height).

If the area is divided into a greater number of strips, then the width of each strip is less and the approximation is more accurate.

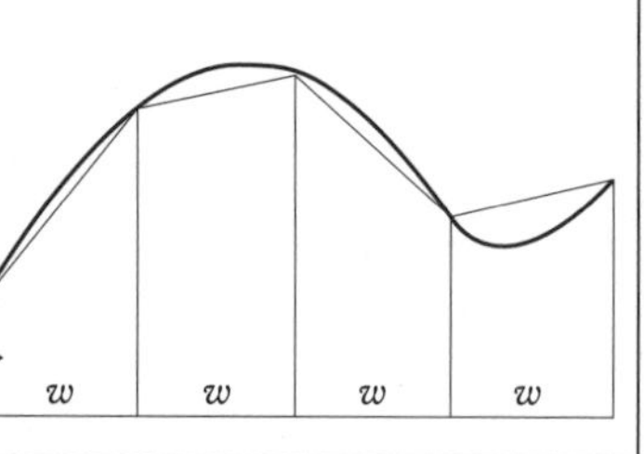

1 Work = force × distance moved (The unit of work is a joule.)

For each of the graphs below calculate the work done by the force in moving from a distance of 5 metres to 10 metres.

(a)

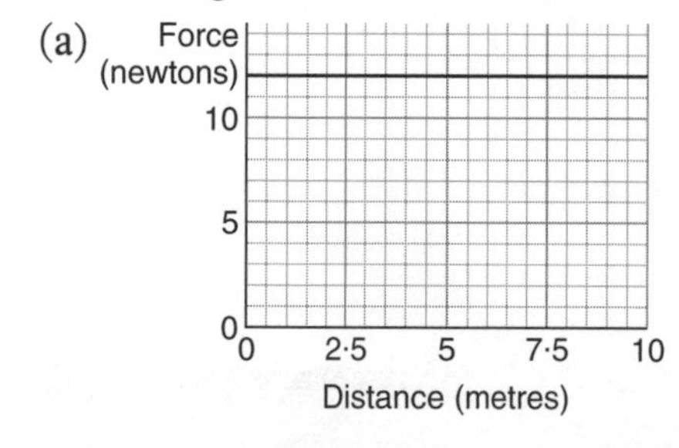

(b)

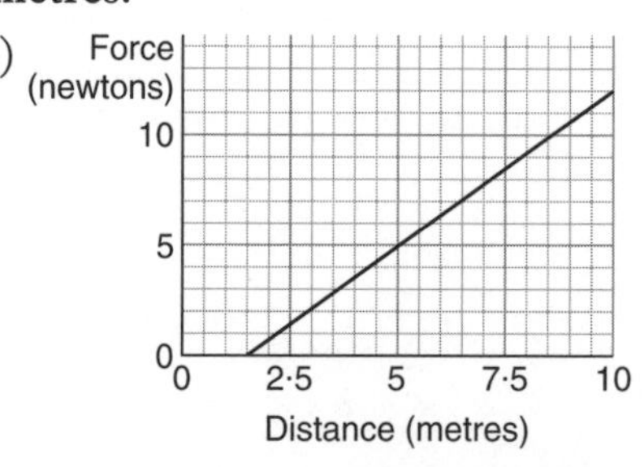

2 The graph shows $f(x) = \sin x$ between $x = 0°$ and $x = 90°$.

(a) Write down the value of y for each of $x = 0°, 15°, 30°, \ldots, 90°$.

(b) Calculate the area of the rectangles under the curve in square units.

(c) Calculate the area of the rectangles above the curve in square units.

(d) Use your answers to (b) and (c) to give an approximation for the area under the graph in square units.

(e) Use the trapezium rule to find another approximation for the area.

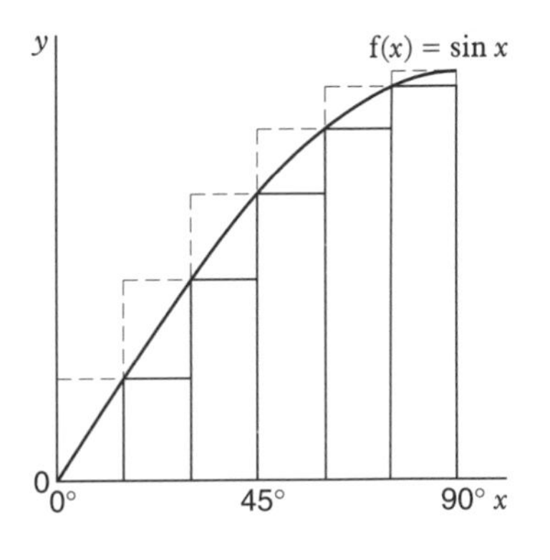

MEG (SMP)

3 This graph shows the cross-section of an aerofoil.

x (cm)	0	20	40	60	80	100
y (cm)	0	12	20	25	26	0

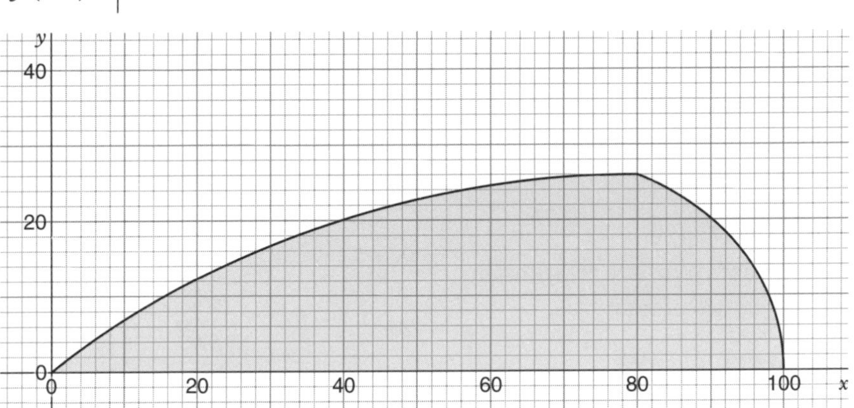

Use the trapezium rule to estimate the shaded area in square centimetres.

MEG (SMP)

4 The areas of several cross-sections of an ancient stone statue of height 12 metres are given in the table.

Distance from base in metres	0	2	4	6	8	10	12
Cross-sectional area in m^2	3	4	4	4	2	2	2

(a) Sketch some possible silhouettes of the statue.

(b) What is the approximate volume of the statue?

(c) If the rock of which the statue is carved has a density of 2900 kg/m^3, find the approximate mass, in tonnes, of the statue.

Answers and hints ► page 117

Fitting functions to data

Lines of best fit

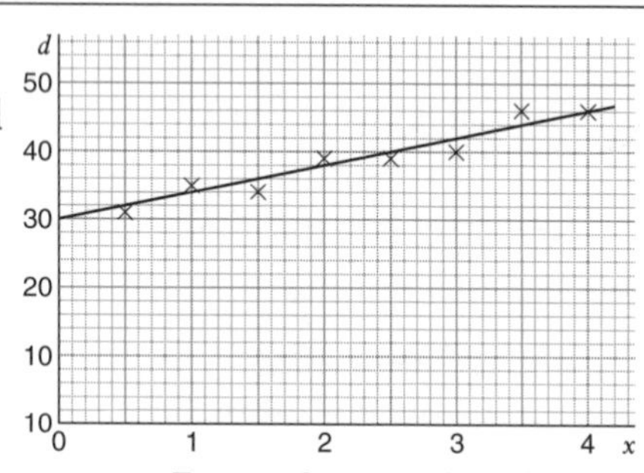

When you have experimental data you often need help to find how the variables are related. Drawing a graph can help. In the diagram eight pairs of experimental data are plotted and the **line of best fit** is drawn. The gradient of the line is 4 and the intercept on the d-axis is 30, so its equation is $d = 4t + 30$. **Straight line graphs ► page 23**

Types of proportionality ► page 12

Non-linear graphs ► page 23

Non-linear functions

Often a set of data suggests that y is proportional to some power of x. Here are the graphs of four types of proportion.

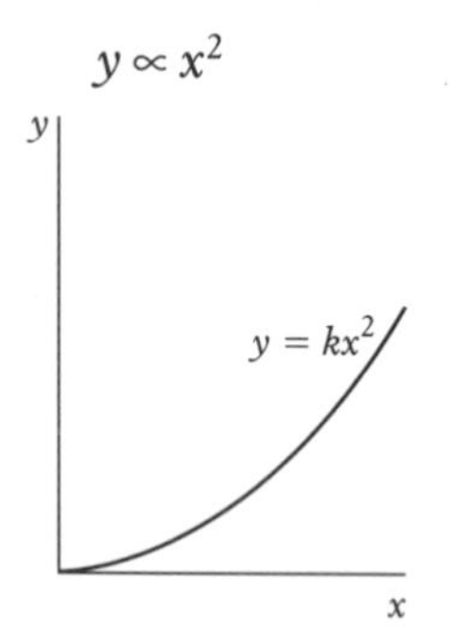

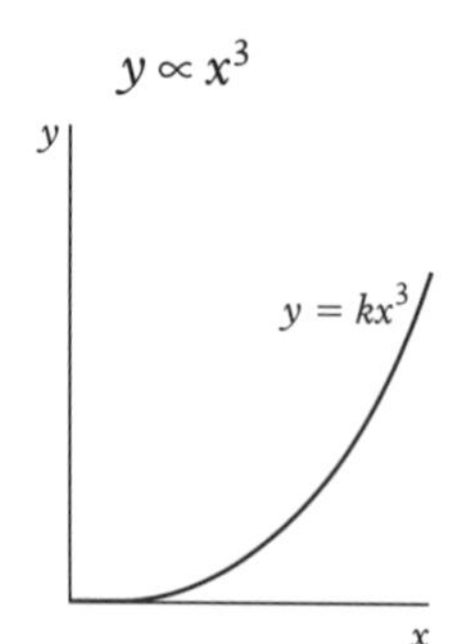

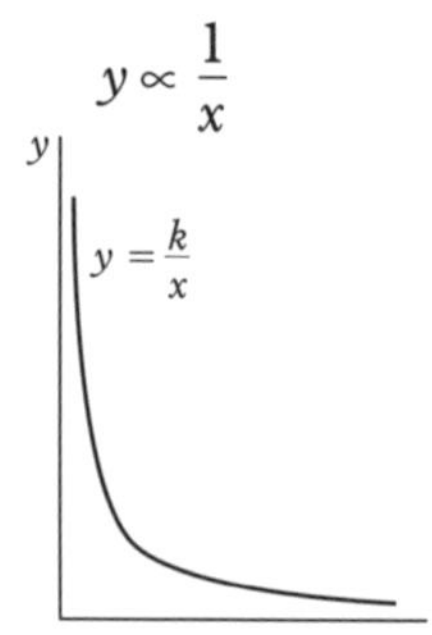

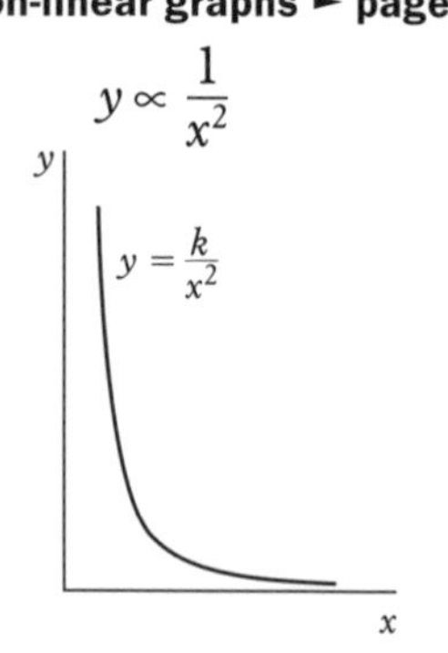

Example

Suppose you want to find the approximate relationship between these experimental values of h and t.

t	2	3·5	4·2	5	5·5
h	15·4	38·5	56·5	78·4	94·5

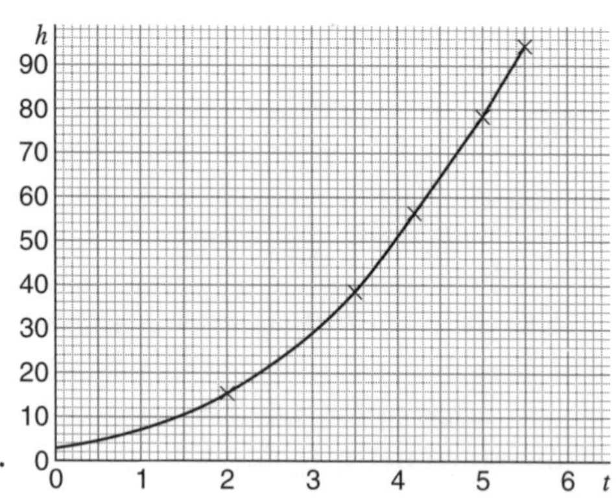

We can tell from the graph that the equation is probably of the form $h = at^2 + b$ or $h = at^3 + b$.

We need to **transform** the graph to linear form to be sure.

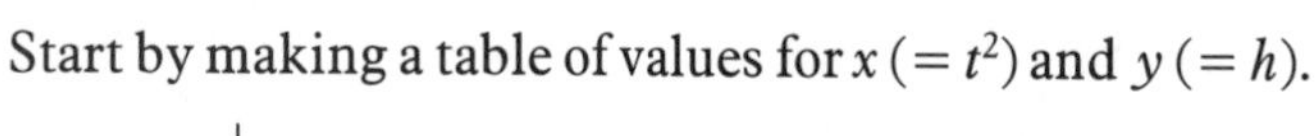

Start by making a table of values for $x\ (= t^2)$ and $y\ (= h)$.

$x(=t^2)$	4	12·25	17·64	25	30·25
$y(=h)$	15·4	38·5	56·5	78·4	94·5

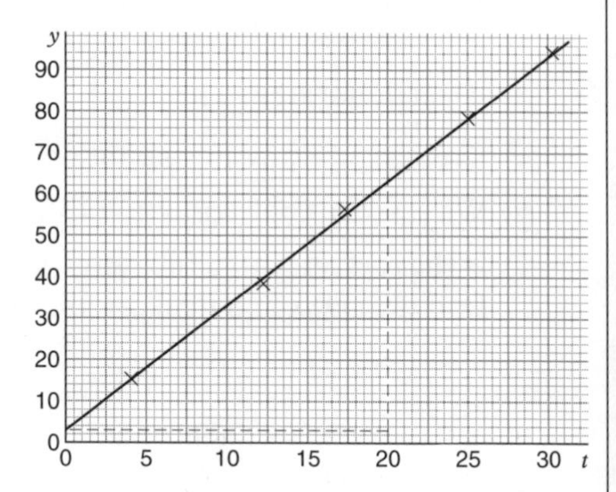

Plot y against x and draw a line of best fit. The gradient of the line is 3 and its intercept on the y-axis is 3, so its equation is $y = 3x + 3$.

But $y = h$ and $x = t^2$, so the approximate relationship is $h = 3t^2 + 3$.

If the relationship between y and x had not been linear you would have had to try $x = t^3$.

1 Heather and John recorded the results of their experiment about the stretching of a spring.

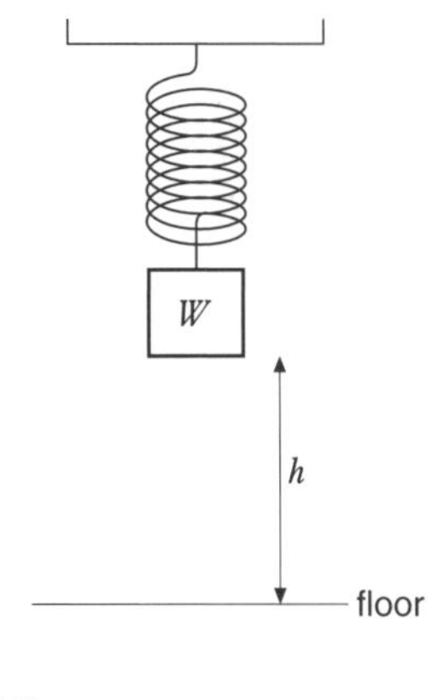

Weight, W (grams)	0	50	100	150	200	250	300
Height above floor, h (cm)	100	87	75	61	48	38	27

Plot their results on a graph and draw in a line of best fit.
Find the equation of your line.

2 Below each graph is a list of equations. From each list select the one which could be the equation of the graph above.

(a)

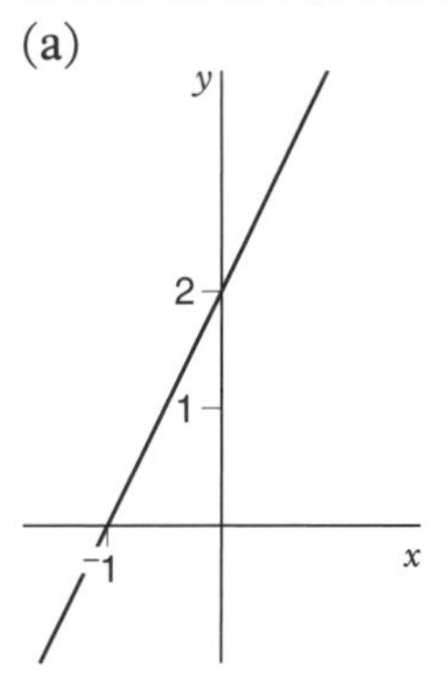

(i) $y = 2x - 2$

(ii) $y = 2x + 2$

(iii) $y = x + 2$

(b)

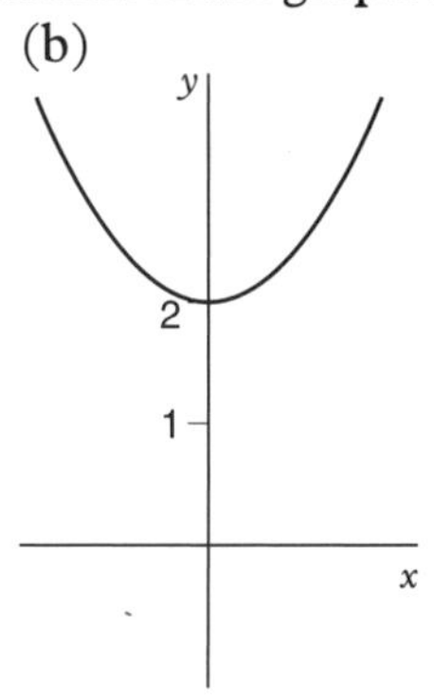

(i) $y = x^2 + 2$

(ii) $y = x^2 - 2$

(iii) $y = x^2 + 2x$

(c)

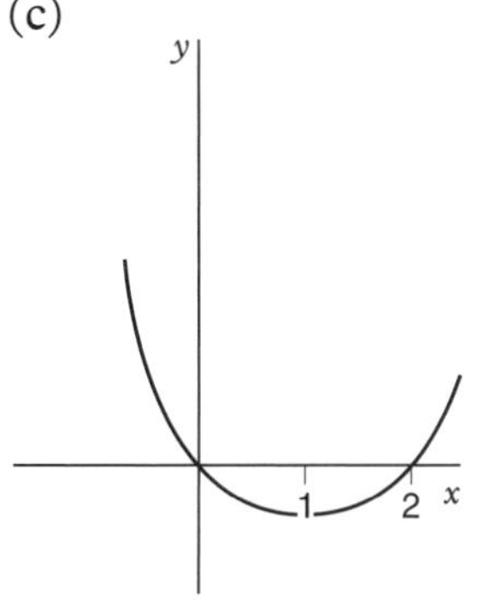

(i) $y = x^2 + 2x$

(ii) $y = x^2 + 2$

(iii) $y = x^2 - 2x$

(d)

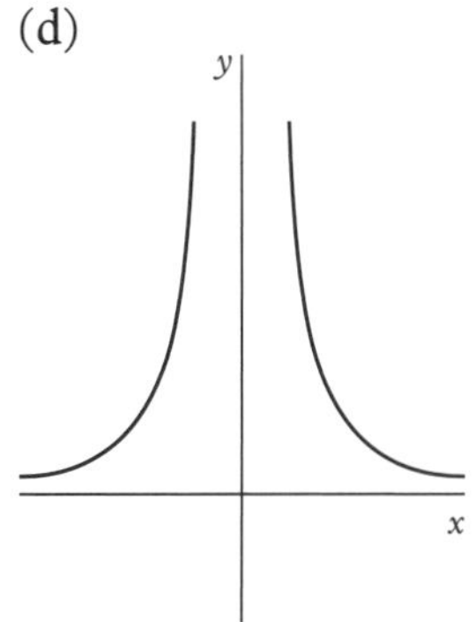

(i) $y = \dfrac{2}{x^2}$

(ii) $y = \dfrac{^-2}{x^2}$

(iii) $y = \dfrac{2}{x}$

NICCEA

3 The costs (£C) of some circular table-cloths are shown in this table.

Diameter of table-cloth (D m)	0·60	0·90	1·50	2·00
Cost (£C)	2·40	3·50	7·10	11·50

(a) Draw the graph of C against D^2.

(b) It is thought that C and D are connected by a formula of the type

$$C = pD^2 + q$$

(i) Does your answer to part (a) support this?

(ii) If it does, estimate the values of p and q. If not, explain why.

MEG

4 (a) Identify which of the following equations could fit this graph.

$y = px + q \qquad y = px^2 + q$

$y = px^3 + q \qquad y = \dfrac{p}{x} + q$

(b) Write down the value of q.

(c) What is the value of p?

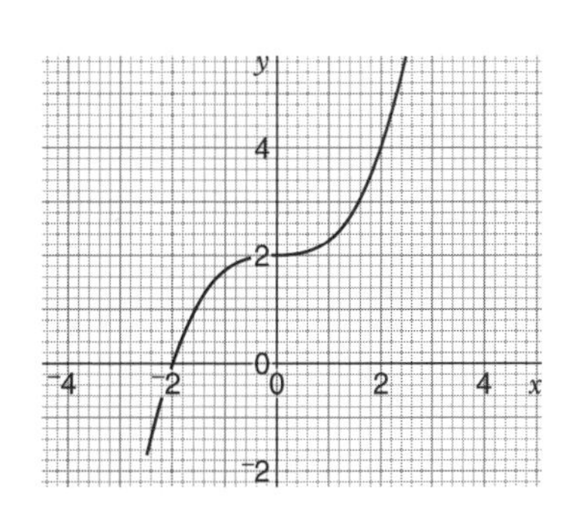

MEG (SMP)

5 It is thought that there is a relationship between N and M of the form $N = \frac{a}{M} + b$. Plot some suitable points, draw the line of best fit and use it to find the values for a and b.

M	0·8	1·2	1·5	1·8	1·9	2·2
N	5·5	4·7	4·3	4·1	4·0	3·8

6 For a given 'light value', the possible combinations of the shutter speed S (expressed as a fraction of a second, so that '60' means '$\frac{1}{60}$s') and the aperture A on a camera are as follows:

Aperture, A	4	5·6	8	11	16	22
Shutter speed, S	60	30	15	8	4	2

Ahab says that S is proportional to the inverse of A and Janine says that S is proportional to the inverse of the square of A.
Show who is probably correct explaining your answer carefully.
Find an equation connecting S and A.

7 (a) The graph (a) is part of $y = A \sin x + k$. What are the values of k and A?
(b) Given that $P = ak^t$, use graph (b) to find the values of a and k.

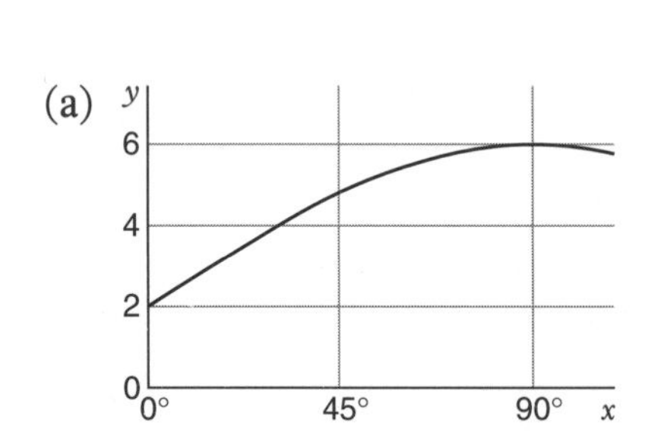

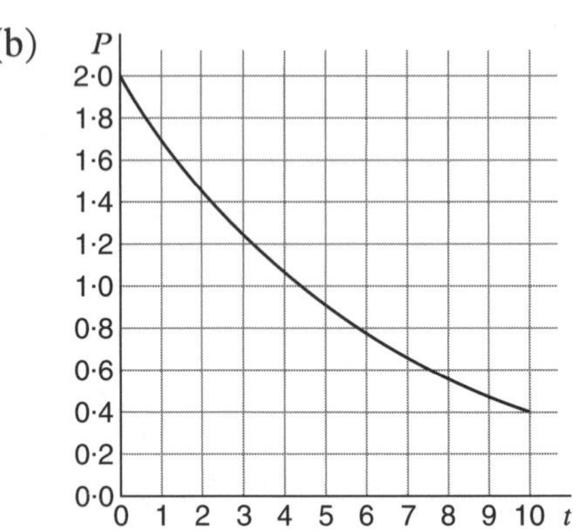

8 The 'maximum safe working load' of a particular type of artificial fibre 'rope' is given as follows.

Diameter when unstretched, d (mm)	5	10	15	20	25
Safe working load, L (newtons)	400	1600	3500	6500	10000

(a) Plot a suitable graph so that the points lie very nearly on a straight line and state a simple relation between the safe working load and the diameter (unstretched) which fits the data well.

The price per metre of the rope depends on the diameter as follows:

Diameter when unstretched, d (mm)	5	10	15	20	25
Price per metre length, P (pence)	10	22	42	70	100

(b) Plot a graph of P against d^2 and deduce a simple equation for P in terms of d.

Answers and hints ► page 117

Mixed algebra

1 Using a scale of one centimetre to represent one unit on each axis draw x-and y-axes, marking each of them from $^{-}2$ to 8.

(a) (i) Mark the point A $(0, {}^{-}1)$ and draw the straight line through it with gradient 2.

(ii) Write down the equation of this line in the form $y = mx + c$.

(b) (i) On the same diagram, mark the point B $(6, 1)$ and draw the straight line through it with gradient $^{-}\frac{1}{2}$.

(ii) Find the equation of this line in the form $y = mx + c$.

(c) Write down the solution to the simultaneous equations which form your answers to (a)(ii) and (b)(ii).

MEG

2 Select from the five graphs one which illustrates each of the following statements.

(a) The time, y, taken for the journey is inversely proportional to the average speed, x.

Inverse proportion ► page 12

(b) The surface area, y, of a sphere is proportional to the square of the radius, x.

(c) The cost, y, of an electricity bill consists of a fixed charge plus an amount proportional to the number of units of electricity used, x.

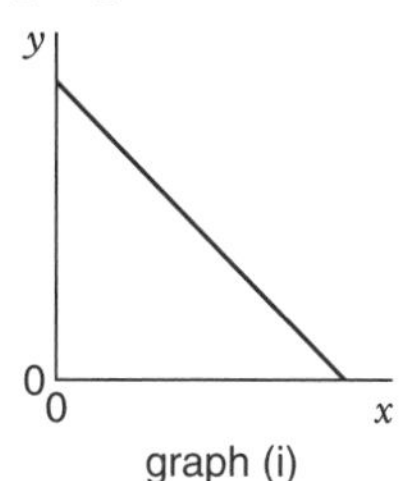

graph (i)

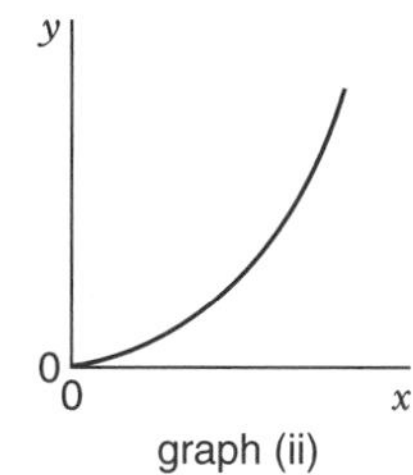

graph (ii)

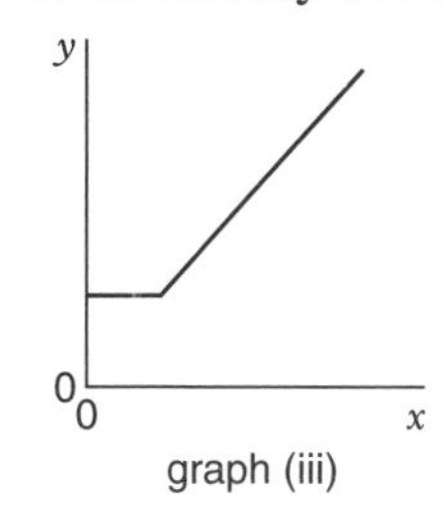

graph (iii)

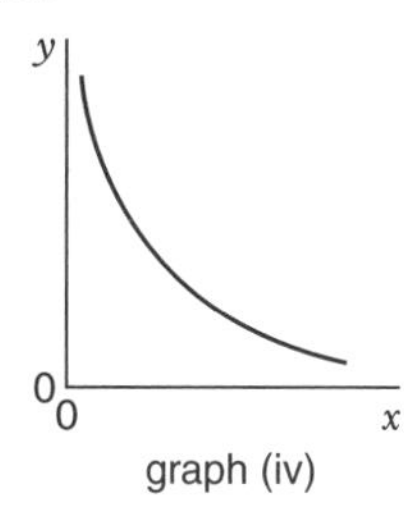

graph (iv)

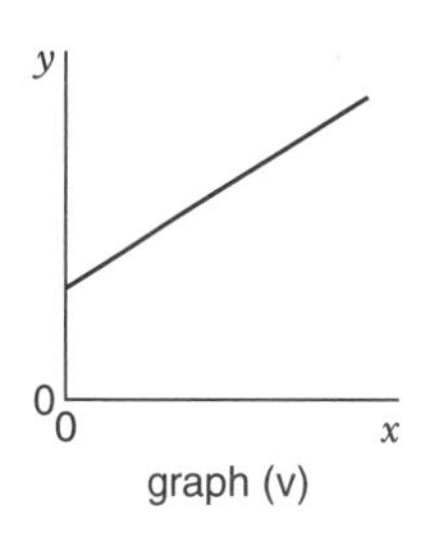

graph (v)

MEG

3 (a) 'A number, x, is chosen so that its reciprocal is two more than the number.' Write this statement as an equation.

(b) Rearrange the equation into a quadratic equation and solve it, giving your answers correct to 2 decimal places.

4 Solve these equations.

(a) $9^{\frac{1}{x}} = 3$ (b) $16^x = 2$ (c) $81^{\frac{1}{4}} = 3^x$ (d) $125^x = \frac{1}{5}$

5 Deena said that '$n^2 - n$ is always divisible by 2 when n is a whole number'.

(a) Factorise the expression $n^2 - n$.

(b) Use your result in (a) to explain whether the statement '$n^2 - n$ is always divisible by 2 when n is a whole number' is true or false.

MEG/ULEAC (SMP)

6 A number n is such that $n^{\frac{1}{3}} = \frac{1}{4}$. Find the values of $n^{\frac{4}{3}}$ and n^{-1}.

7 A straight line, $y = mx + c$, passes through the points P and Q which have coordinates (7, 1) and ($^-3$, $^-4$) respectively.

(a) Write down two equations connecting m and c.

(b) Solve the two equations to find the values of m and c.

(c) The x-coordinate of a third point, R, on the same straight line is 12. Calculate the y-coordinate of R.

WJEC

8 R, S, and T are related by the formula $\frac{1}{R} + \frac{1}{S} = \frac{1}{T}$.

(a) Find T when $R = 32$ and $S = {}^-50$.

(b) Find T when $S = 3\frac{1}{4}$ and $R = 6\frac{1}{2}$. Give your answer as a fraction.

MEG/ULEAC (SMP)

9 (a) Draw the graph of $y = x^2 - 2x$ for values of x between $^-2$ and 4.

(b) By drawing a suitable straight line, estimate the solutions to the equation $x^2 - 4x + 2 = 0$.

(c) By drawing another straight line on the same diagram, solve the inequality $x^2 - 2x \geq 2$.

10 Sebastian sets out from home to catch the bus to school.
He jogs the first 300 metres at a steady speed of x metres per second.
He runs the next 210 metres at a steady speed of $(x + 4)$ metres per second.

(a) Obtain an expression, in terms of x, for
(i) the time in seconds he takes to jog the 300 m,
(ii) the time in seconds he takes to run the 210 m.

(b) Write down and simplify an expression for the total time he takes.

(c) (i) The total time Sebastian takes is 130 seconds.
Form an equation in x and show that it simplifies to $13x^2 + x - 120 = 0$.
(ii) Solve this equation.

MEG (SMP)

11 A firework is launched vertically upwards at time $t = 0$.
Its vertical height, h, in metres from its point of launch is given by $h = 20t - 5t^2$ where t is in seconds.

(a) Use a graphical method to estimate
(i) the first time at which the firework reaches a height of 12 metres,
(ii) the maximum height reached by the firework.

(b) Use your graph to estimate the speed of the firework when it first reaches a height of 12 metres.

(c) Solve an appropriate quadratic equation to calculate the second time that the firework reaches a height of 12 metres.
Give your answer correct to 2 decimal places.

SEG

Questions 12, 13 and 14 are on worksheets H3 and H4.

Answers and hints ► page 119

SHAPE, SPACE AND MEASURES

Angles

Angle properties

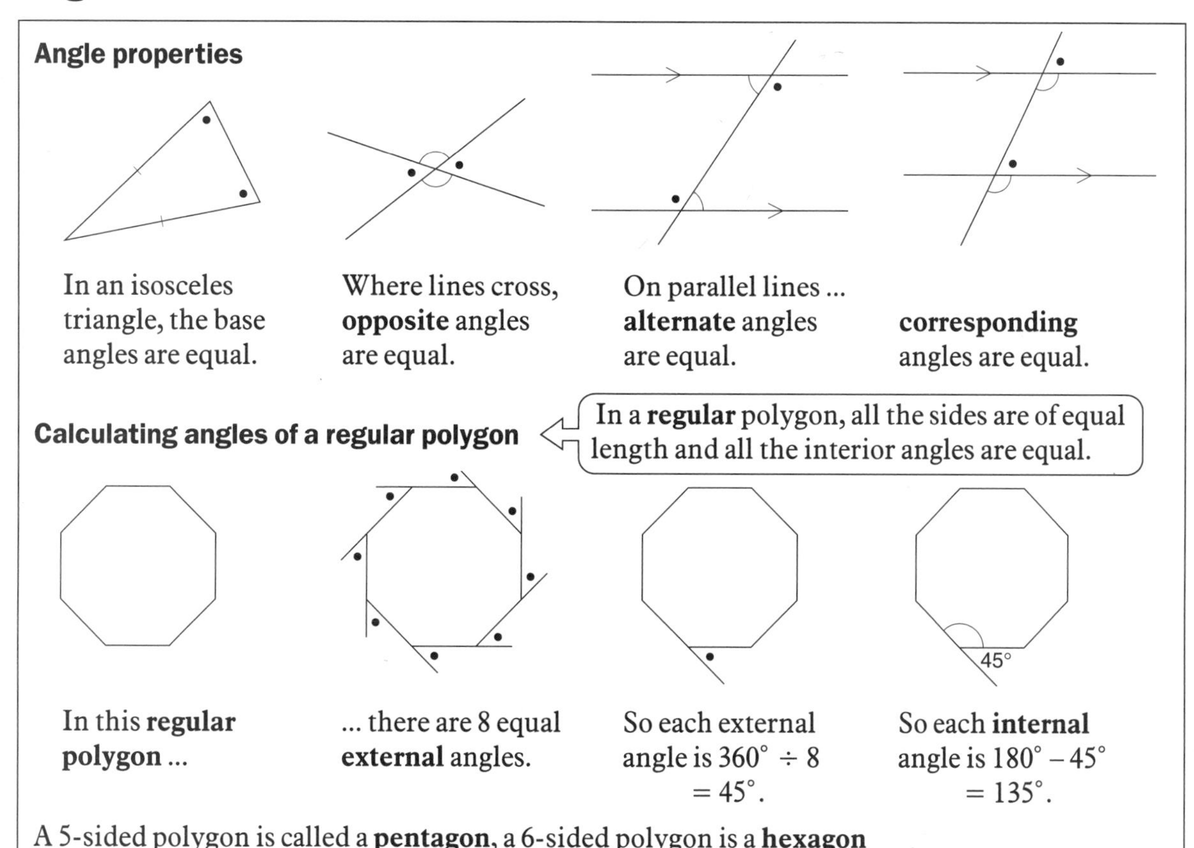

In an isosceles triangle, the base angles are equal.

Where lines cross, **opposite** angles are equal.

On parallel lines ... **alternate** angles are equal.

corresponding angles are equal.

Calculating angles of a regular polygon

In a **regular** polygon, all the sides are of equal length and all the interior angles are equal.

In this **regular polygon** ...

... there are 8 equal **external** angles.

So each external angle is $360° \div 8 = 45°$.

So each **internal** angle is $180° - 45° = 135°$.

A 5-sided polygon is called a **pentagon**, a 6-sided polygon is a **hexagon** and an 8-sided polygon is an **octagon**.

Congruence

Two figures are said to be **congruent** if they are equal in all respects.
For example, in a triangle, when three angles and three sides in one triangle are equal to the *corresponding* three angles and three sides in the other triangle.

Two triangles are congruent if any one of these sets of conditions is satisfied:

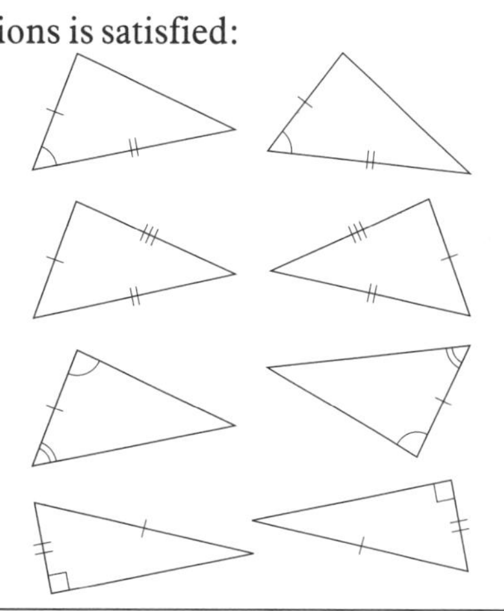

- Two sides of one triangle are equal to two sides of the other and the angle between these sides (the included angle) is equal (**SAS**).
- The three sides of one triangle are equal to the three sides of the other triangle (**SSS**).
- Two angles and a side of one triangle are equal to the angles and the corresponding side of the other (**ASA**).
- Each triangle is right-angled and the hypotenuse and one side of one triangle are equal to the hypotenuse and a side of the other triangle (**RHS**).

Angle properties of a circle

You need to know the the following facts about angles in circles.
(In each diagram, the angles with dots in them are equal.)

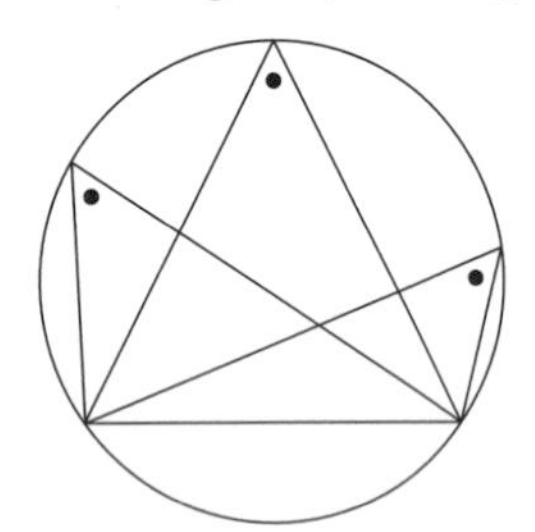

Angles in the same segment are equal.

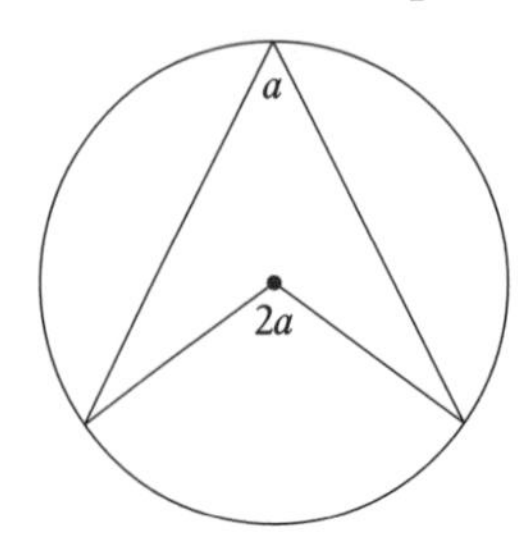

The angle at the centre of a circle is twice that at the circumference.

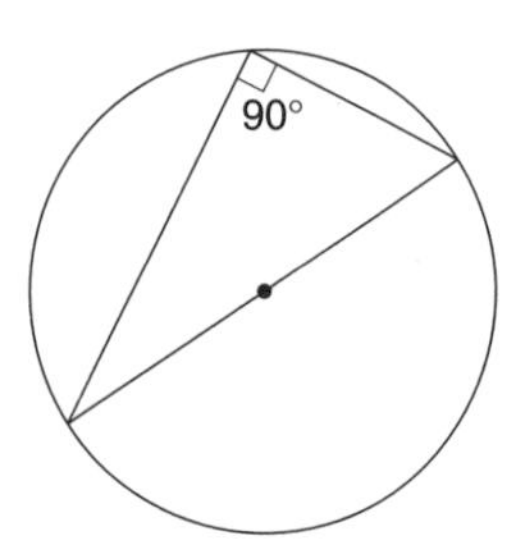

The angle in a semicircle is 90°.

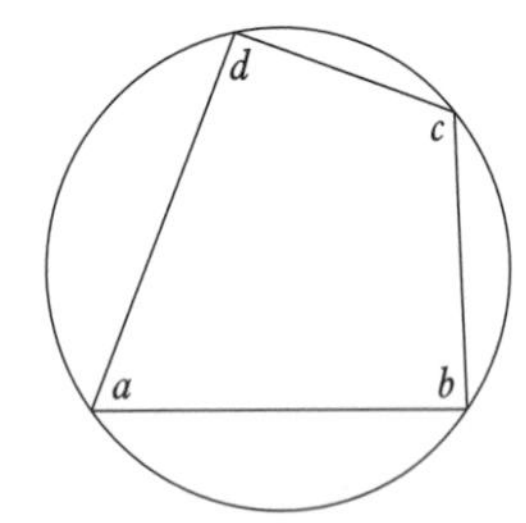

A **cyclic quadrilateral** has all its vertices on a circle. Its opposite angles add up to 180°. ($a + c = 180°, b + d = 180°$)

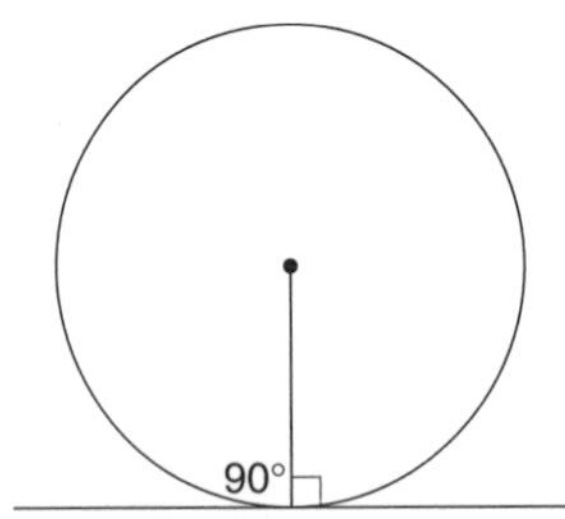

The angle between a tangent and the radius where it touches the circle is 90°.

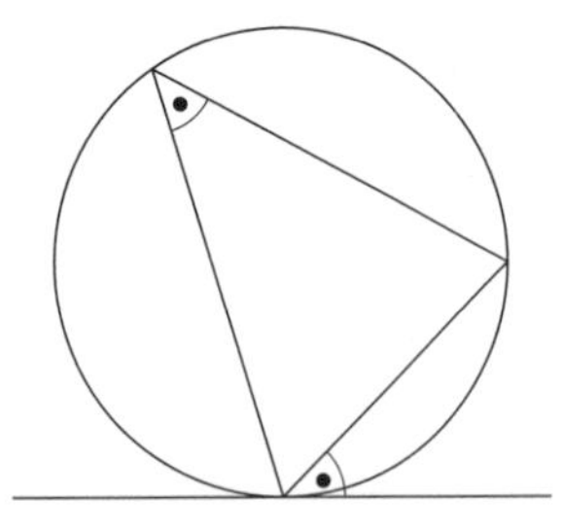

The angle between a tangent and any chord where it touches the circle is equal to the angle in the alternate segment.

Tangents from a point are equal in length. (AB = AC)

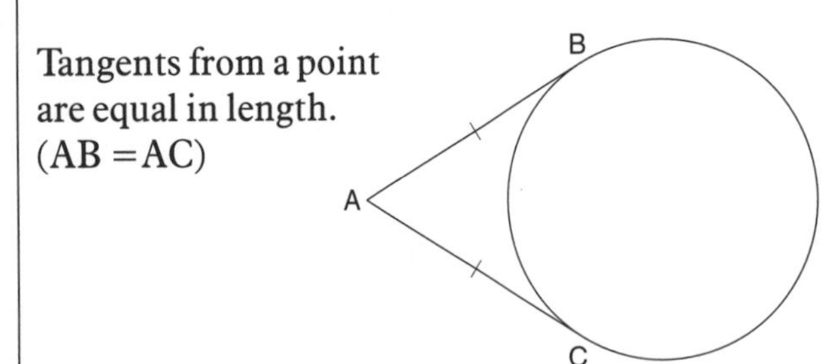

You may be asked to use any of the above properties to prove results. You may need other facts about angles as well.

It is a good idea to mark the angles you can work out on a copy of the diagram.

When you are asked to give reasons in a proof, you should give a reason like 'Tangents from a point' or 'Angle in the alternate segment'.

1 The diagram shows a regular pentagon.
Some of the vertices are joined up.

(a) Calculate the sizes of angles (i) PQR (ii) PRQ.

(b) Name two congruent triangles in the diagram.

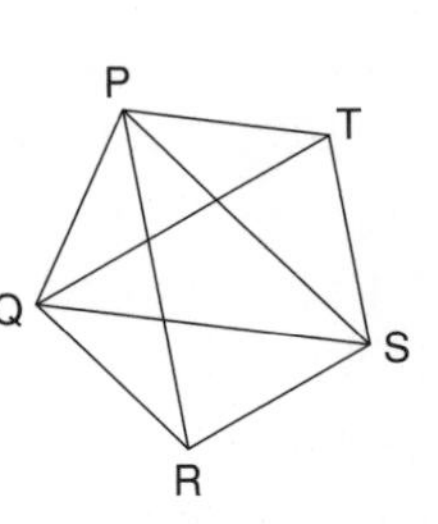

2 Two interior angles of a polygon are 155° and 165°.
Each of the remaining angles is 170°.
How many sides has the polygon?

MEG

None of the diagrams is to scale.

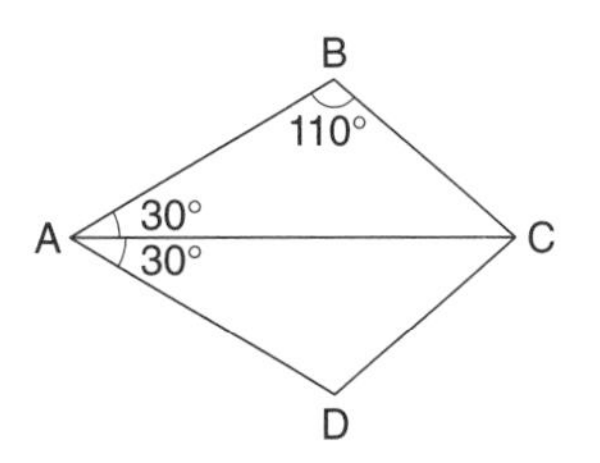

3 In quadrilateral ABCD, AB = AD.
Calculate the sizes of angles ADC and BCD.

4 The diagram shows a circle, centre O.
PR and PQ are tangents to the circle at R and Q.
Showing your working clearly

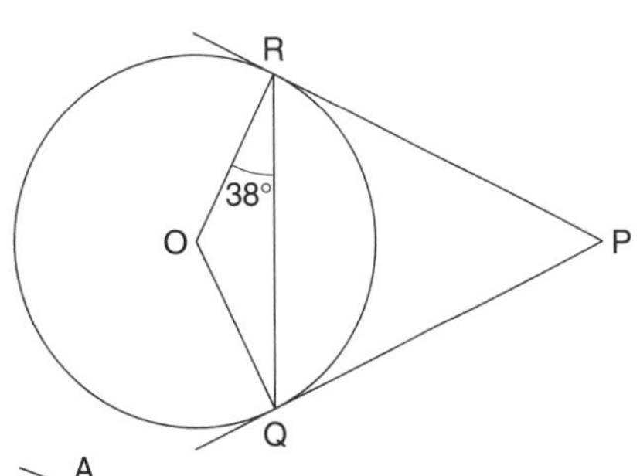

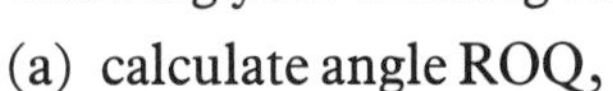

(a) calculate angle ROQ,

(b) calculate angle RQP,

(c) calculate angle RPQ.

MEG

5 AD and BD are tangents to the circle.
O is the centre of the circle.

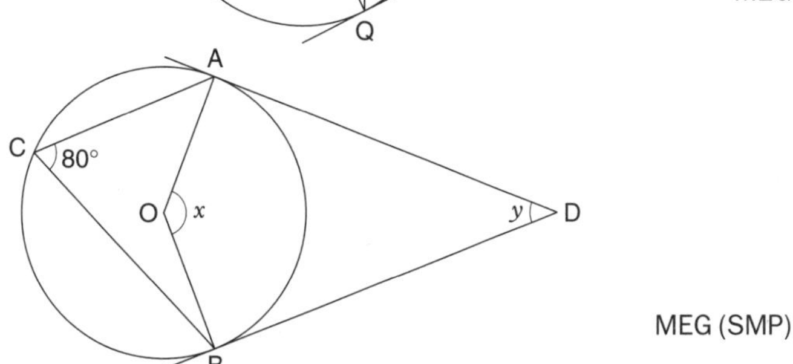

(a) Calculate the value of x.
Give a reason for your answer.

(b) Calculate the value of y.
Give a reason for your answer.

MEG (SMP)

6 ABCD are four points on the circumference of a circle, centre O.
BOD is a straight line.
Given that $\angle BDC = 43°$ and $\angle ABD = 62°$, calculate the following angles, giving reasons for your answers.

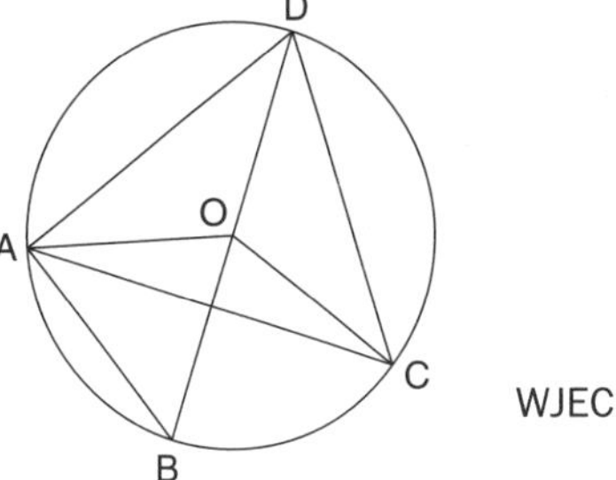

(a) $\angle BAC$ (b) $\angle DAC$

(c) $\angle AOD$ (d) $\angle OAC$

WJEC

7 LMNP is a cyclic quadrilateral, centre O.
LM is equal to MN and angle LON is 160°.
Calculate the following angles, giving reasons for your answers.

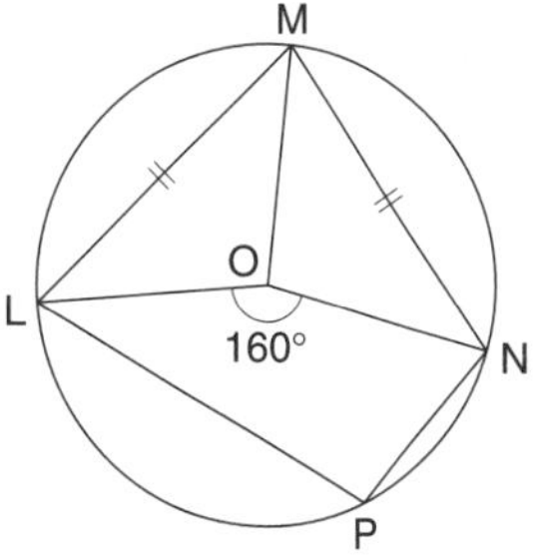

(a) $\angle LPN$ (b) $\angle LMO$

(c) $\angle OLM$ (d) $\angle MOL$

8 In the diagram, the straight line ABC is a tangent to the circle with centre at O.
The circle passes through the points B, D, E and F.
BOE is a diameter of the circle and AFE is a straight line. $\angle BEF = 42°$.
Find the sizes of the angles marked v, w, x, y and z.

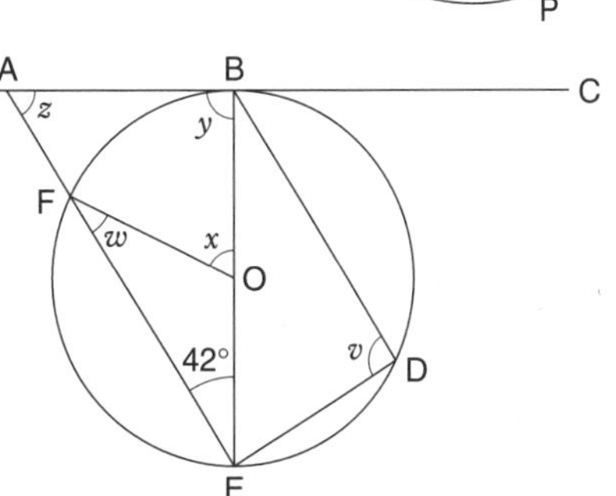

WJEC

9 (a) In the diagram, O is the centre of the circle and AD is a diameter. Angle BAO is p and angle CAO is q.

(i) **Without using angle properties of circles,** find the size of angle BOC in terms of p and q.

(ii) What is the relationship between angle BAC and angle BOC?

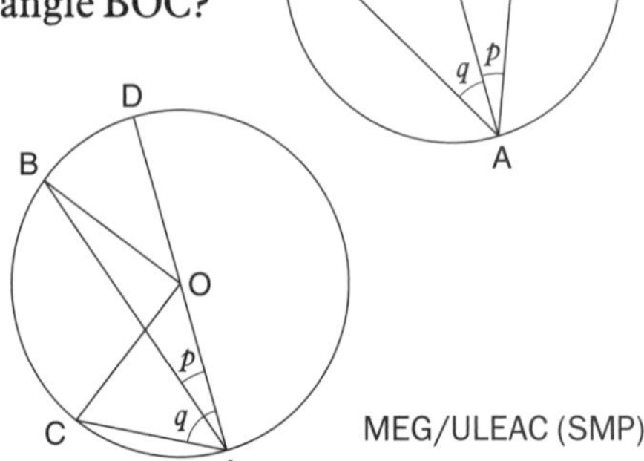

(b) In this diagram, O remains the centre of the circle and AD is a diameter. Angle BAO is again p, and angle CAO is q.

Show that the relationship between angle BAC and angle BOC is still true.

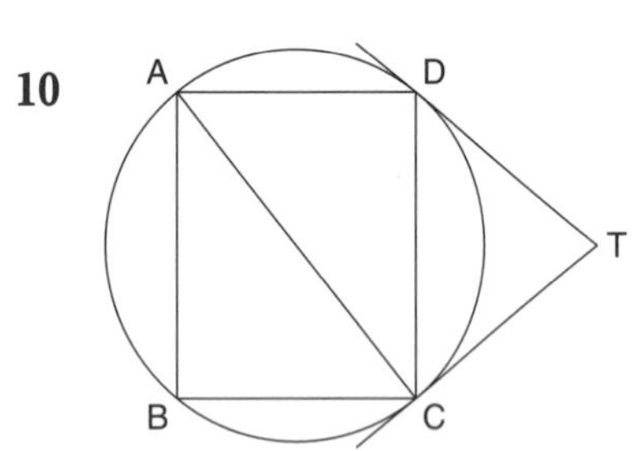

MEG/ULEAC (SMP)

10

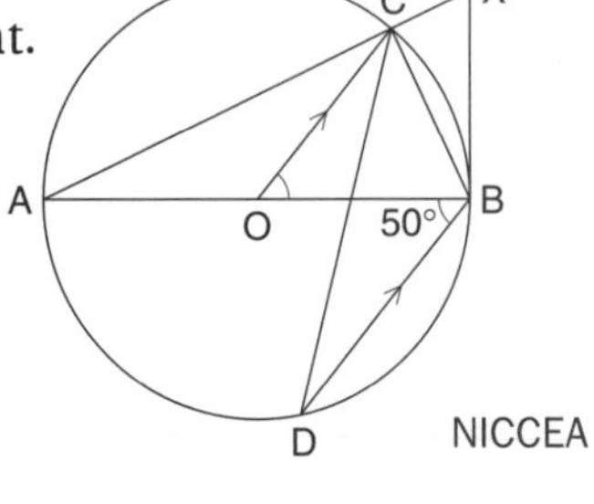

The diagram shows the circle through the vertices A, B, C and D of a rectangle.

(a) Explain why AC is a diameter of the circle.

(b) Tangents to the circle at D and C meet at T. Angle DCT = 42°.

(i) Giving a reason for each step of your working, calculate angle CTD.

(ii) Giving a reason for each step of your working, calculate angle CAB.

MEG

11 (a) O is the centre of the circle, AB is a diameter and BX is a tangent. The lines OC and DB are parallel. Angle ABD = 50°.
Find the sizes of:

(i) angle COB (ii) angle OBC (iii) angle CBX
(iv) angle BXC (v) angle CAB

(b) Prove that OC bisects angle ACD.
State your reasons clearly.

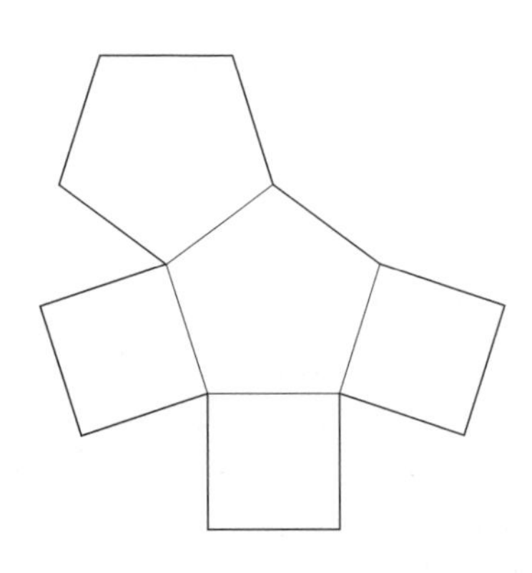

NICCEA

12 (a) Calculate the interior angle of a regular pentagon.

(b) What is the interior angle of a square?

(c) Use your answers to (a) and (b) to deduce (with reasons) whether it is possible to draw a tessellation using only squares and pentagons.

Answers and hints ► page 121

Pythagoras and trigonometry in two dimensions

In any **right-angled** triangle you should be able to use Pythagoras' rule (see the formula sheet) and the three trigonometric ratios:

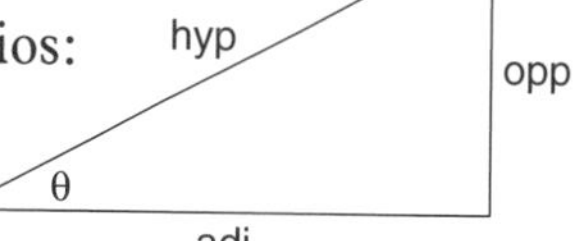

$$\sin\theta = \frac{\text{opposite}}{\text{hypotenuse}} \qquad \cos\theta = \frac{\text{adjacent}}{\text{hypotenuse}} \qquad \tan\theta = \frac{\text{opposite}}{\text{adjacent}}$$

Pythagoras' rule and the trig ratios are special cases of the **sine** and **cosine** rules. You use these rules in triangles that have no right-angle.

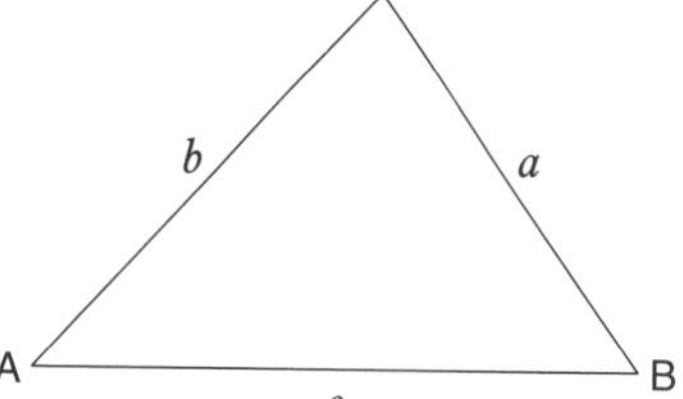

Sine rule $\dfrac{a}{\sin A} = \dfrac{b}{\sin B} = \dfrac{c}{\sin C}$

Cosine rule $a^2 = b^2 + c^2 - 2bc\cos A$ or $\cos A = \dfrac{b^2 + c^2 - a^2}{2bc}$

Which rule to use when solving a triangle

If you know a side and the opposite angle, use the sine rule.
(Remember that if you know two angles of a triangle, it is easy to find the third!)
Otherwise use the cosine rule.

Area of a triangle

For any triangle, the area is $\frac{1}{2}ab\sin C$.

Example

In triangle PQR, find the length of PQ.
Calculate the area of the triangle.

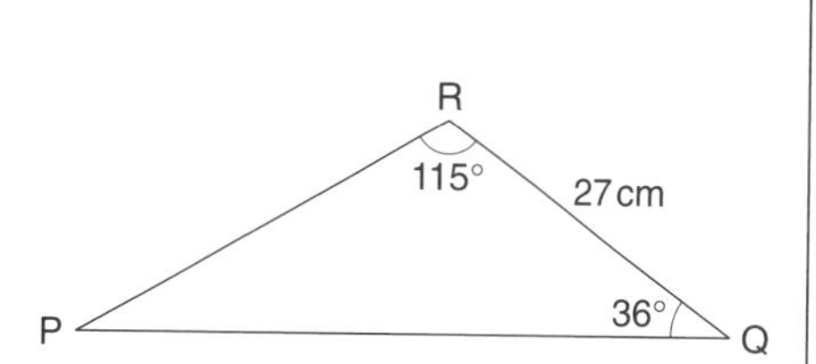

Here we don't seem to know a side and opposite angle. But if you try to use the cosine rule, you need to know two sides – which we don't.

However, we can easily work out angle P $= 180° - 36° - 115° = 29°$.

PQ is opposite angle R; suppose it is r cm long.

We know that $\dfrac{r}{\sin 115°} = \dfrac{27}{\sin 29°}$, so $r = \dfrac{27}{\sin 29°} \times \sin 115° = 50{\cdot}474\ldots$

So PQ = 50 cm (to 2 s.f.).

To find the area we need two sides and the angle between them.
Now we have found PQ, we know PQ, RQ and the angle Q.
So the area of the triangle

$= \frac{1}{2} \times 27 \times 50{\cdot}474\ldots \times \sin 36°\,\text{cm}^2$ ⇐ *Use the uncorrected value of PQ.*

$= 400{\cdot}516\ldots\,\text{cm}^2$

$= 400\,\text{cm}^2$ (to 2 s.f.)

None of the diagrams is to scale.

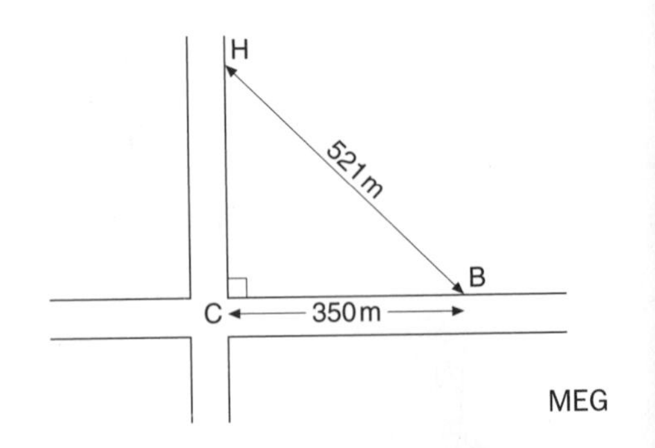

1 Mohamed takes a short cut from his home (H) to the bus stop (B) along a footpath HB.
How much further would it be for Mohamed to walk to the bus stop by going from H to the corner (C) and then from C to B?
Give your answer correct to the nearest metre.

MEG

2

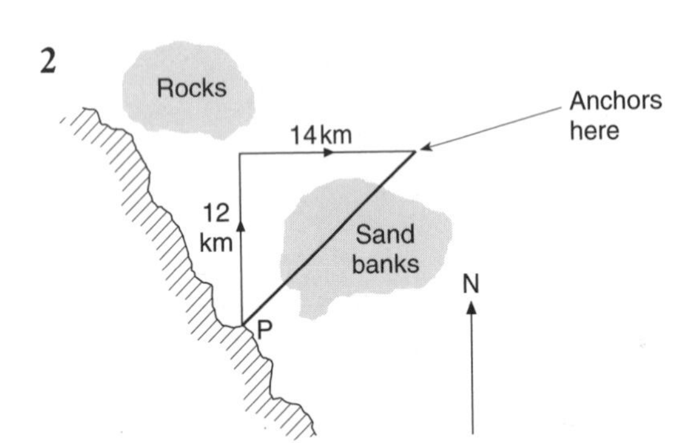

A ship leaves port P and sails north for 12 km. It then changes course and travels east for 14 km before anchoring.

(a) How far away from P is the ship anchored?

(b) Calculate the bearing of the ship from P when it is anchored.

MEG/ULEAC (SMP)

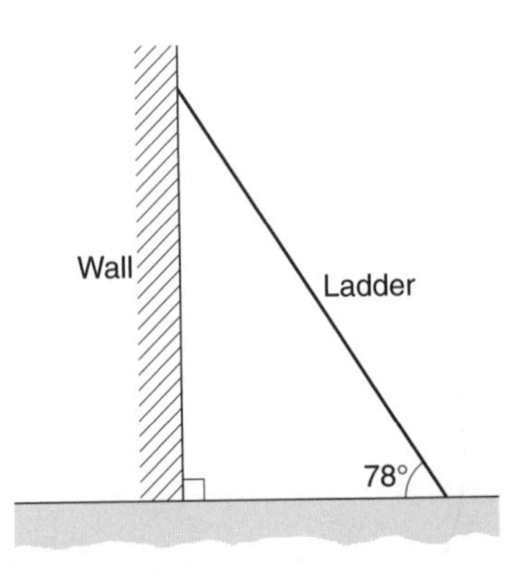

3 The diagram shows a ladder 6 m long leaning against a house wall.

(a) Calculate the height of the top of the ladder from the ground.

The ladder is moved so that the foot of the ladder is 1·6 m from the wall.

(b) What angle with the ground does the ladder make now, as it leans against the wall?

MEG/ULEAC (SMP)

4

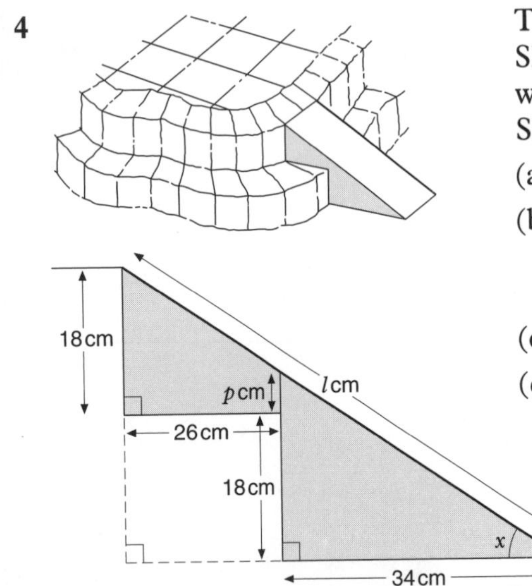

Two steps lead up to the patio in Cathy's garden. She wants to build a wooden ramp so that she can wheel heavy loads up on to the patio.
She decides to make it as shown on the left.

(a) Calculate the sloping length of the ramp, l cm.

(b) Calculate the area of wood required to make one side of the ramp (the area shown shaded in the diagram).

(c) Calculate the angle of inclination of the ramp, x.

(d) For extra strength, Cathy decides to fix a support on the edge of the ramp. Calculate the height of this support, shown as p cm on the diagram.

MEG (SMP)

5 A and B are two checkpoints on an orienteering course.
Mala runs 800 m due north from the start, S, to A.
She knows that B is on a bearing of 247° from S and on a bearing of 205° from A.

(a) Calculate angle ABS.

(b) Mala runs in a straight line from A to B.
Calculate the distance that she runs from A to B.

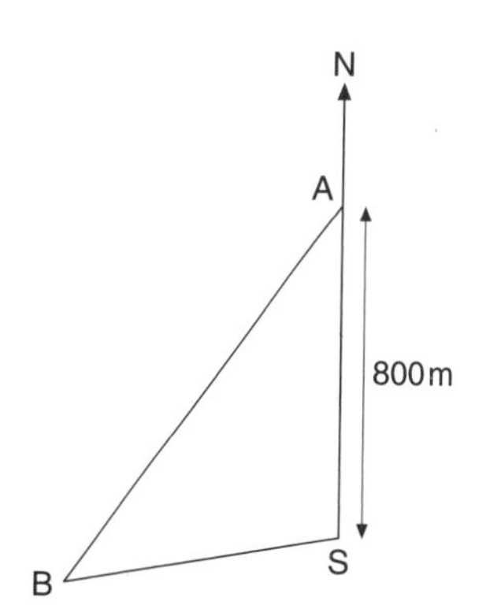

MEG

6

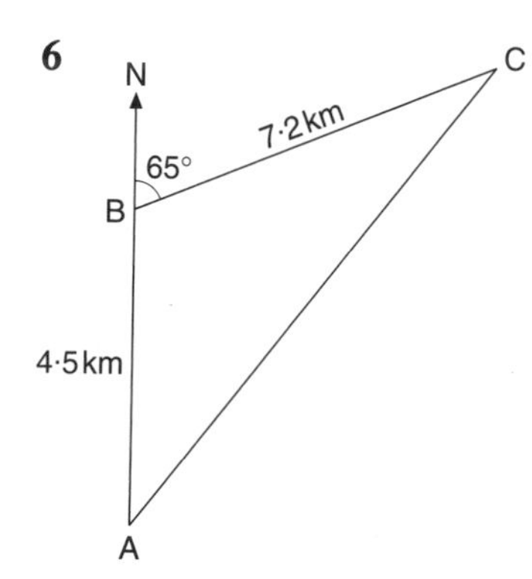

A sailing race has a triangular course, ABC, where AB = 4·5 km and BC = 7·2 km. B is due north of A and C is on a bearing N65°E (065°) from B.
Calculate

(a) the length of CA, the final part of the race,

(b) the bearing on which the boats should sail to get from C to A.

WJEC

7 Jeff rents a triangular field PQR alongside a straight road.
The side PQ of the field along the road is 120 m long.
The angles in the corners P and Q of the field are shown.

(a) Jeff needs to put new fencing along the two sides away from the road.
What length of fencing does he need?

(b) What is the area of Jeff's field in square metres?

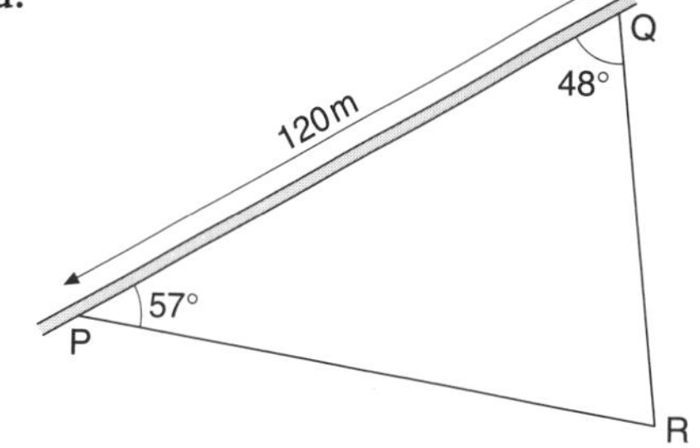

8 A trawler is allocated a triangular area of sea OXY in which she is allowed to fish.
OX is defined as 85 km long on a bearing of 274°.
OY is defined as 65 km long on a bearing of 310°.

(a) Calculate the area of sea in which the trawler is allowed to fish.

(b) The trawler sails to X and sets a course from X to Y. Calculate the bearing on which she must sail from X to Y.

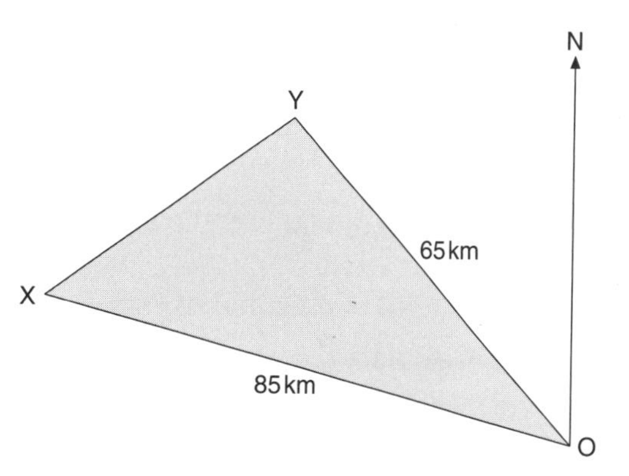

Answers and hints ► page 123

Pythagoras and trigonometry in three dimensions

Pythagoras in three dimensions

Problems in three dimensions can usually be broken down into two-dimensional parts.
But if you have to find the diagonal length of a cuboid (l in the diagram), you will find it useful to know that

$$l^2 = a^2 + b^2 + c^2$$

$QR^2 = a^2 + c^2$
$QP^2 = QR^2 + b^2 = a^2 + b^2 + c^2$

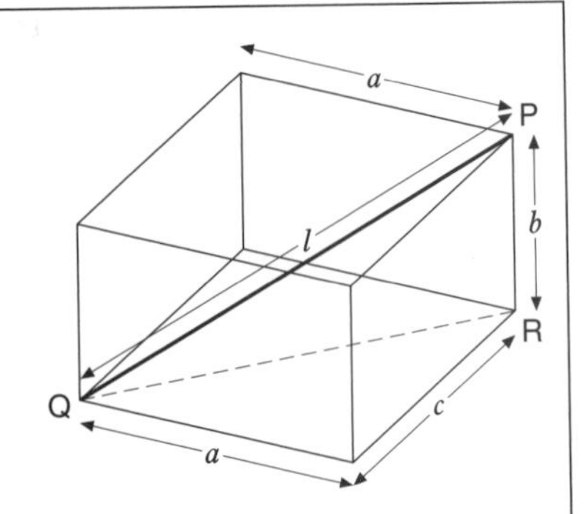

Angles in three dimensions

To find the angle between a line and a plane, you need to find a triangle which is perpendicular to the plane and has the line as one of its sides.

Example

The diagram shows a triangular prism. Angle FCB = 90°.
Work out

(a) the length of EB,

(b) the angle EB makes with the plane DEFC.

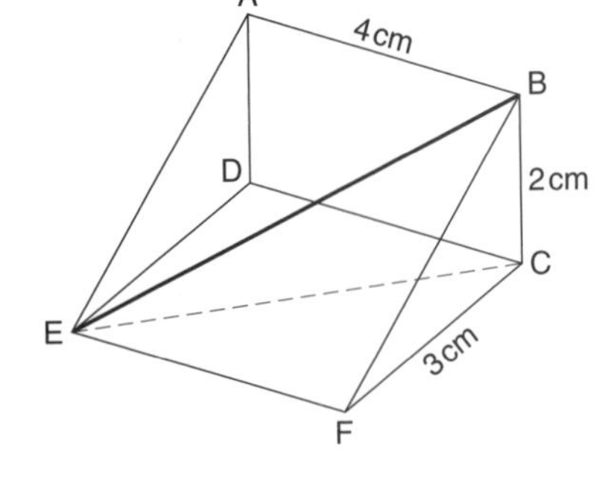

(a) $EC^2 = EF^2 + FC^2 = 4^2 + 3^2 = 16 + 9 = 25$
$EC = \sqrt{25} = 5$
Now triangle ECB is right-angled.
So $EB^2 = EC^2 + CB^2 = 5^2 + 2^2 = 25 + 4 = 29$.
$EB = \sqrt{29} = 5{\cdot}385\ldots = 5{\cdot}4$ cm (to 2 s.f.)

(Note that you could go straight to this answer, since EB is the diagonal of a cuboid with sides 2, 3 and 4.
So $EB^2 = 2^2 + 3^2 + 4^2 = 4 + 9 + 16 = 29$, $EB = \sqrt{29} = 5{\cdot}4$ cm (to 2 s.f.)

(b) The angle between the line EB and the plane DEFC is ∠BEC, since triangle BEC is perpendicular to the plane.

$\angle ECB = 90°$, so $\sin BEC = \dfrac{BC}{EB} = \dfrac{2}{\sqrt{29}}$

$\angle BEC = 21{\cdot}801\ldots° = 22°$ (to 2 s.f.)

Always use unrounded answers if you need to use a length or angle again.

1 In the rectangular box shown, AB is 60 cm and BC is 25 cm.
The length of diagonal AG is 72 cm.

(a) Calculate the length of AC.

(b) Calculate the height of the box.

(c) Calculate the angle that AG makes with the face ADHE.

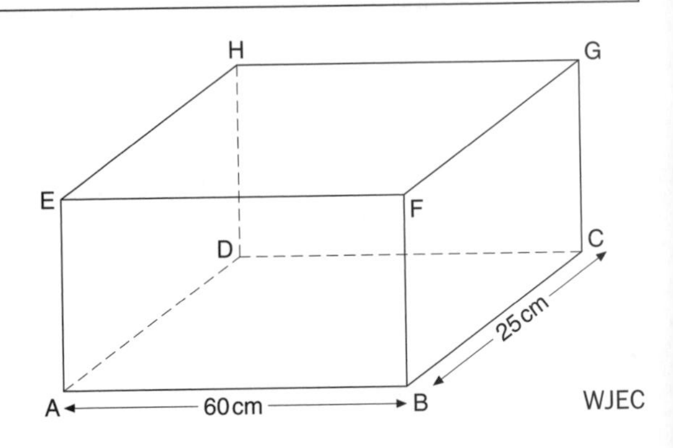

WJEC

2 A telegraph pole, CT, stands vertically upright in one corner of a flat triangular field, ABC, as shown in the diagram.
The field is right-angled at C, ∠ABC is 35° and the length of side AB is 47 m.
The angle of elevation of T from A is 18°.
Calculate

(a) the length of CA,

(b) the height of the telegraph pole.

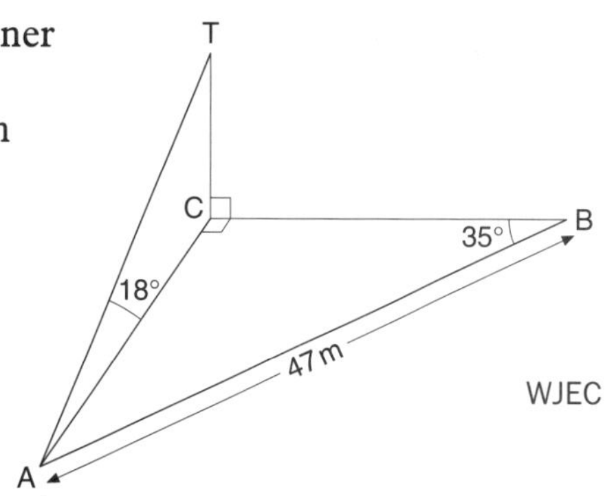

WJEC

3 Peter and Queenie are surveying on a horizontal plane PQT, using PQ as a base line.
TM is a television mast of height 70 m. Angle PTQ = 134°.
The angles of elevation of M from P and Q are 32° and 35° respectively.

(a) Show that the distance from P to Q is approximately 195 m.

(b) Calculate the size of angle TPQ.

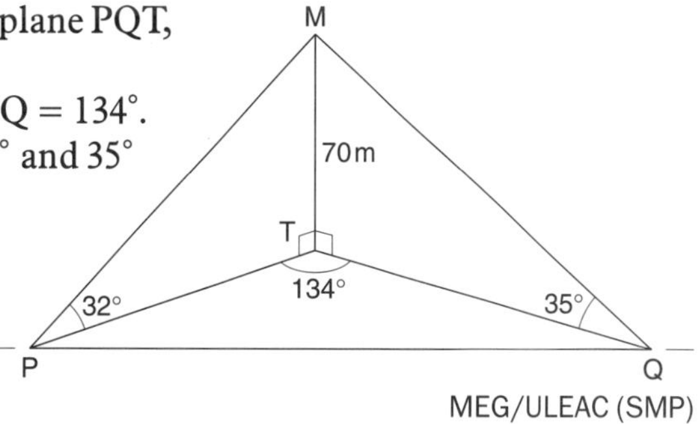

MEG/ULEAC (SMP)

4 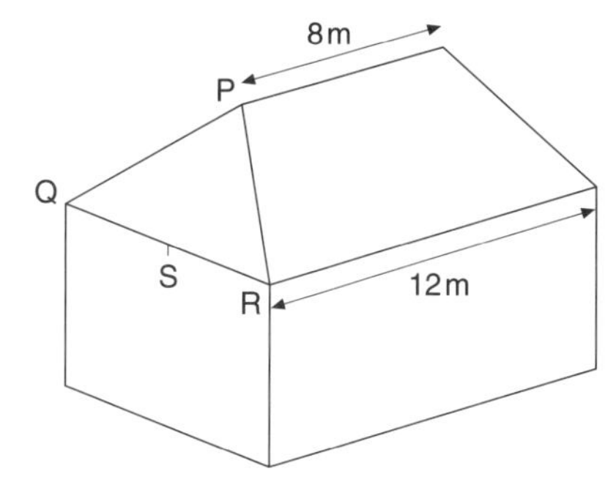

The roof of a house has a top ridge of length 8 m as shown.
The length of the house is 12 m.
The height of the ridge above the base of the roof is 1·8 m.
All of the sloping edges of the roof are the same length as each other. S is the midpoint of the edge QR.
Calculate the angle that PS makes with the horizontal.

MEG/ULEAC (SMP)

5 The frame of a rotary garden clothes line consists of a vertical pole, AB, with four identical arms CD, CE, CF and CG each hinged to AB at a point C. The point C is 0·7 m below B. The arms are inclined at 50° to the vertical and their ends D, E, F and G lie in the same plane as B so that DEFG is a square.

(a) Calculate the length of each arm.

(b) (i) Calculate DB.

(ii) Part of the clothes line joins the points D, E, F and G. Calculate the total length of this part of the clothes line.

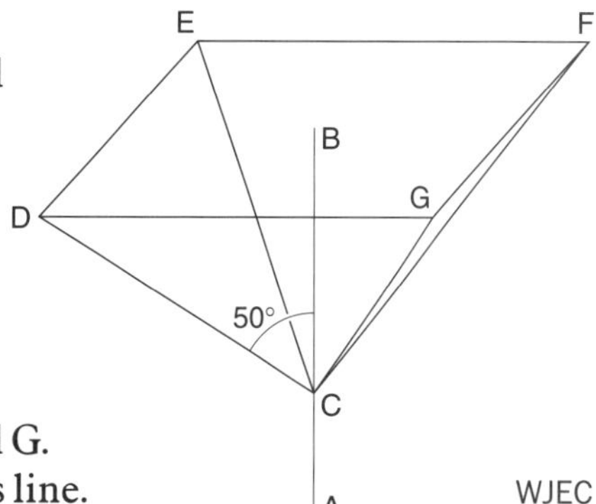

WJEC

Answers and hints ► page 125

Length, area and volume 1

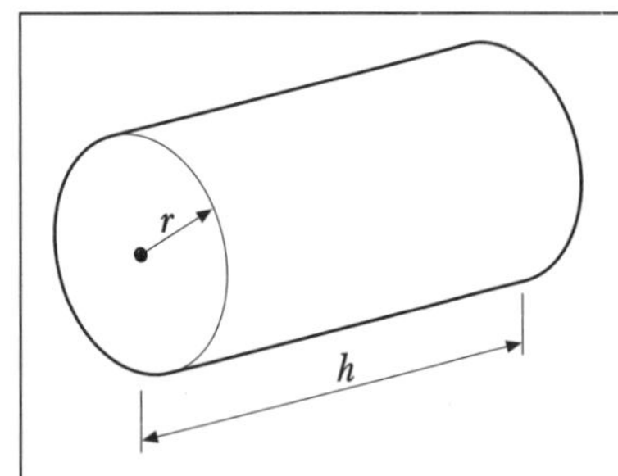

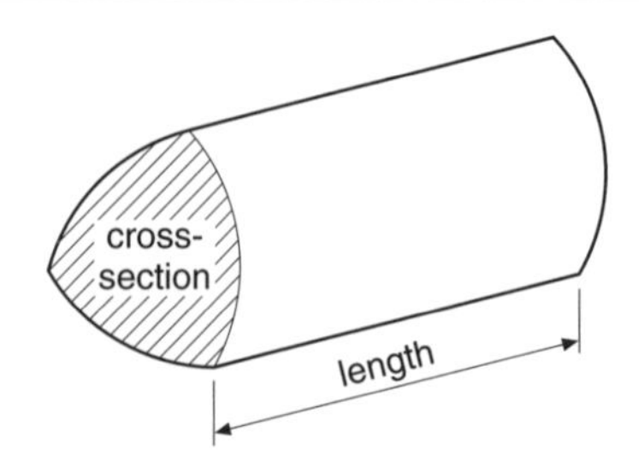

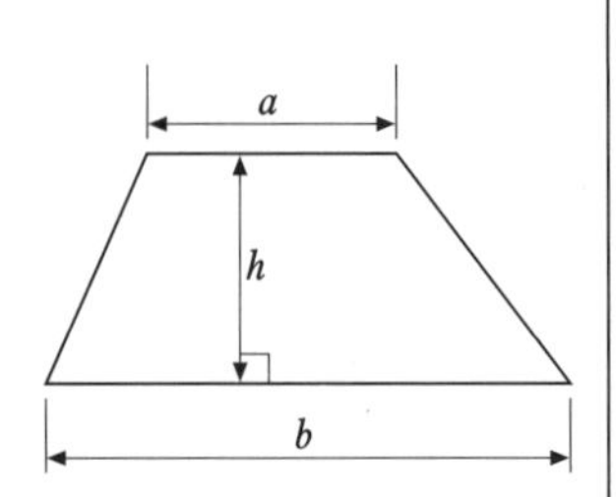

Volume of a cylinder = $\pi r^2 h$
Curved surface of a cylinder = $2\pi rh$

Volume of a prism
= area of cross-section × length

Area of a trapezium
= $\frac{1}{2}(a + b)h$

One way to check formulas is to consider **dimensions**.
The dimension of perimeters is always **length**.
The dimension of areas is **length²**, and the dimension of volumes is **length³**.

Example
Could the formula for the volume of a cone be $2\pi rh$?

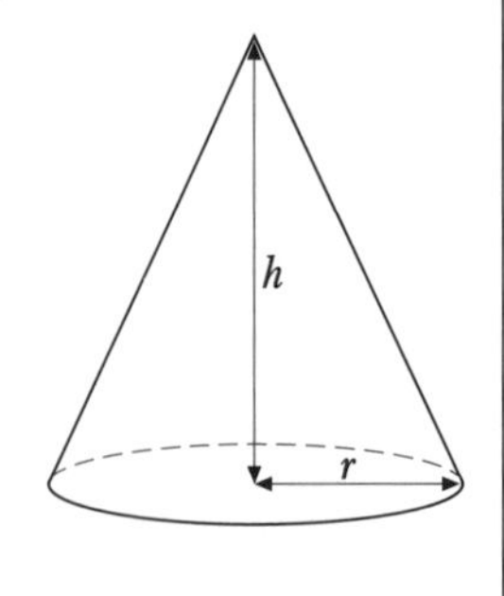

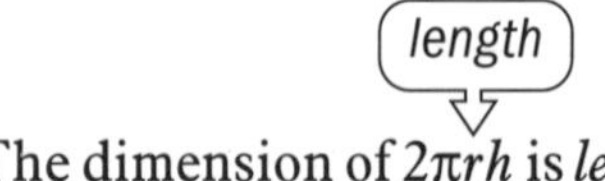

The dimension of $2\pi rh$ is *length* × *length* = *length*².

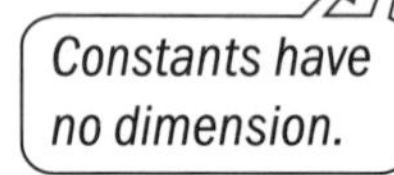

A volume must have dimension *length*³, so the formula cannot be correct.
(The actual formula is $\frac{1}{3}\pi r^2 h$.)

1 A garden cloche is made by spreading a sheet of polythene over wire hoops. The ends are not covered.

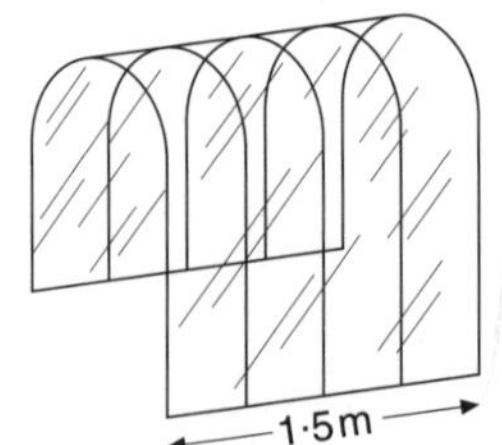

The measurements of each hoop are shown in the diagram. The top of the hoop is a semicircle.

(a) Calculate the length of wire to make one hoop, including the part below the ground.

(b) What area of polythene is needed to make a cloche 1·5 m long if the hoops are covered to ground level?

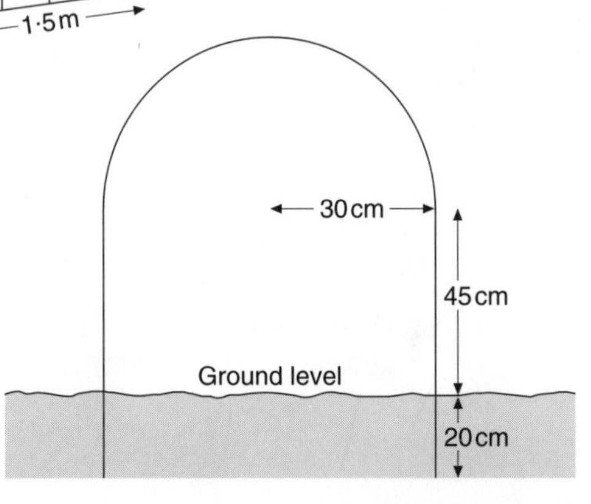

MEG/ULEAC (SMP)

2 A potter is designing novelty mugs for a client.
These are two of her designs.

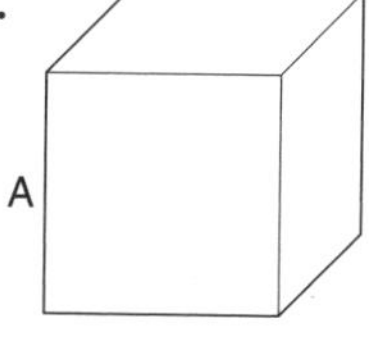

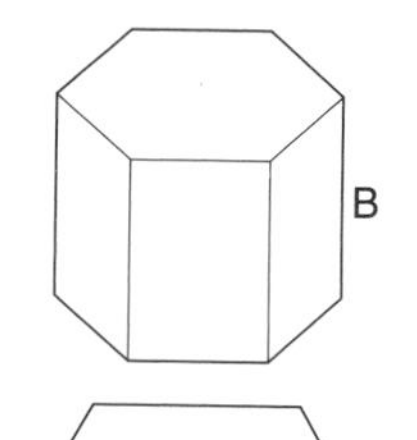

These sketches show the dimensions of each cross-section.
A mug has to hold 0·45 litres.

(a) Calculate the height of each mug.

(b) The client decides that he would prefer a cylindrical mug 10 cm high.
What would its radius be?

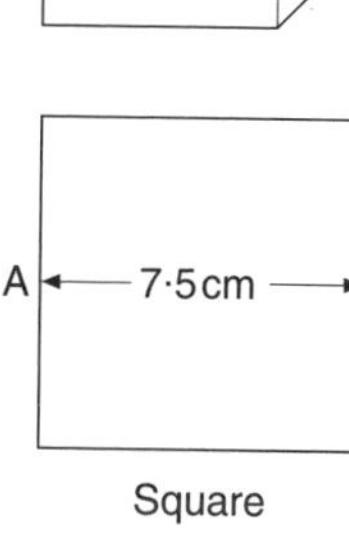

Square

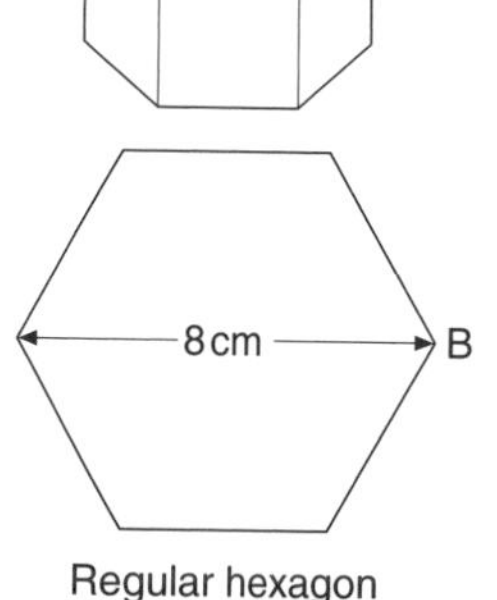

Regular hexagon

MEG/ULEAC (SMP)

3 The spare leads for my pencil are in a case which is a prism.
The cross section is a rhombus.
The diagonals of the rhombus measure 8 mm and 13 mm.
The length of the case is 71 mm.
Calculate its volume.

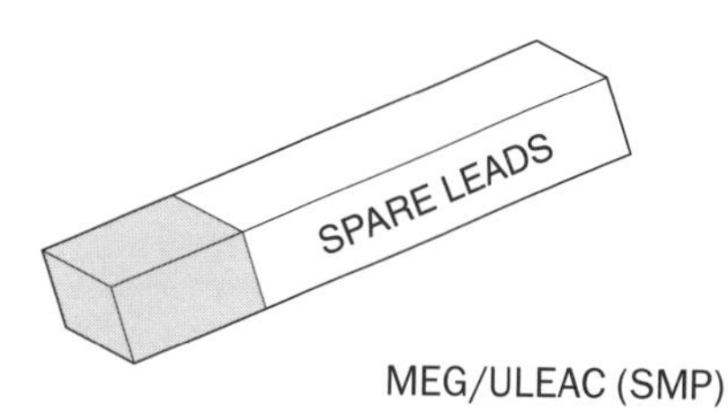

MEG/ULEAC (SMP)

4 ABCD is a trapezium where AD = 10 cm,
BC = 5 cm and the perpendicular height is 7·1 cm.
AB is equal to CD.
All four sides are tangents to the circle.

(a) Calculate the area of the circle.

(b) Calculate the total area of the regions shaded on the diagram.

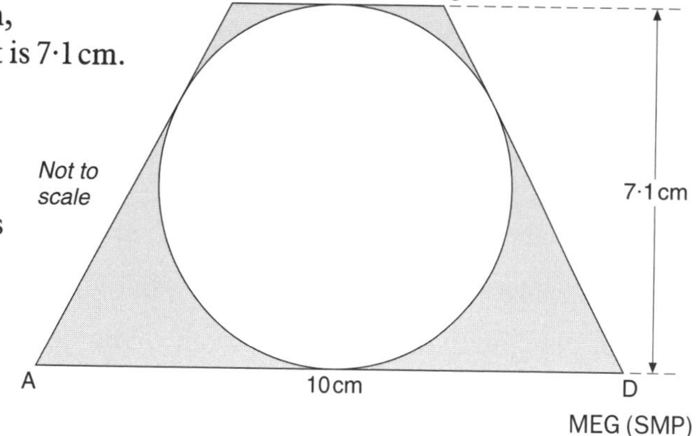

MEG (SMP)

5 A wok is part of a sphere of radius r.
Which one of the following expressions could be correct for the surface area of the wok?

$\frac{4}{3}\pi r^3$ $\quad$ $\frac{2}{7}\pi\sqrt{r}$ $\quad$ $3{\cdot}6\pi r$ $\quad$ $\frac{4}{7}\pi r^2$

MEG (SMP)

Answers and hints ► page 127

Length, area and volume 2

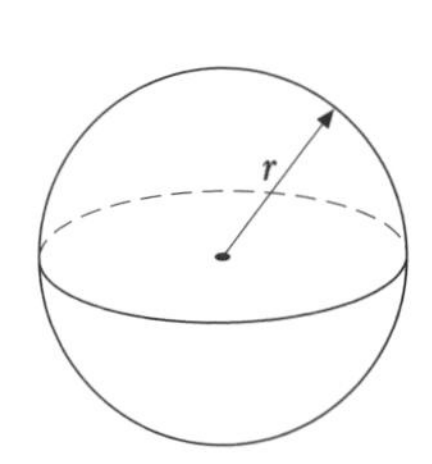

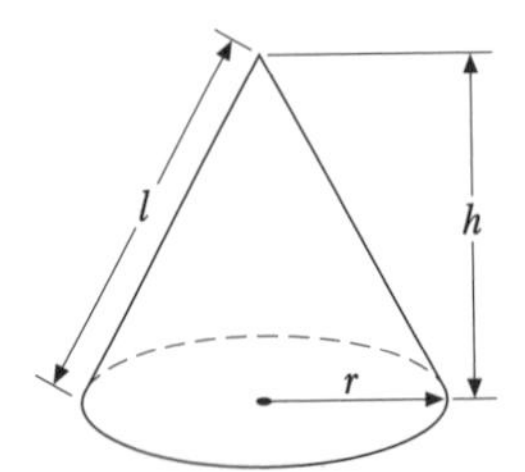

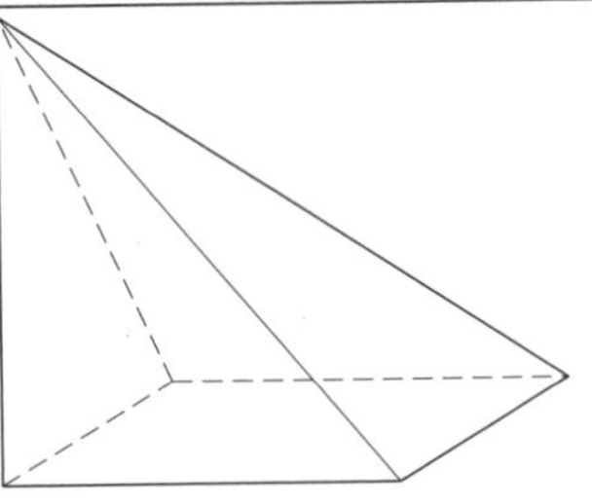

Volume of a sphere = $\frac{4}{3}\pi r^3$

Surface area of a sphere = $4\pi r^2$

Volume of a cone = $\frac{1}{3}\pi r^2 h$

Curved surface area of a cone = πrl

Volume of a pyramid = $\frac{1}{3}$ base area × height (The pyramid need not be regular.)

Arcs and sectors

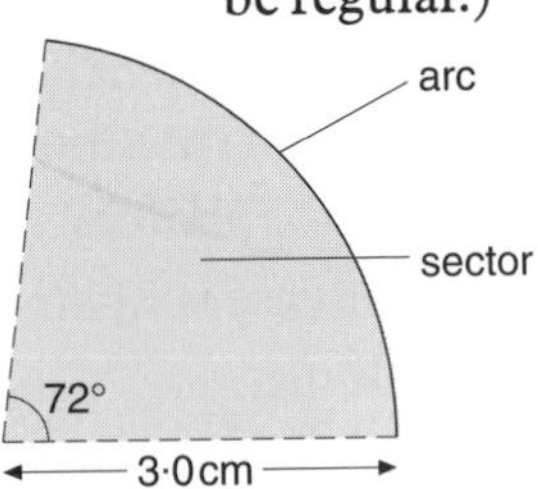

To work out the length of an arc of a circle, think of it as part of a whole circle.

This arc is $\frac{72}{360}$ths of the whole circle.

The circle has radius 3 cm, so its circumference is $2\pi \times 3\text{ cm} = 6\pi\text{ cm} = 18{\cdot}84\ldots\text{ cm}$.

So the length of the arc = $\frac{72}{360} \times 18{\cdot}84\ldots\text{ cm} = 3{\cdot}769\ldots\text{ cm} = 3{\cdot}8\text{ cm}$ (to 2 s.f.).

You work out the area of a sector using fractions in a similar way.

Segments

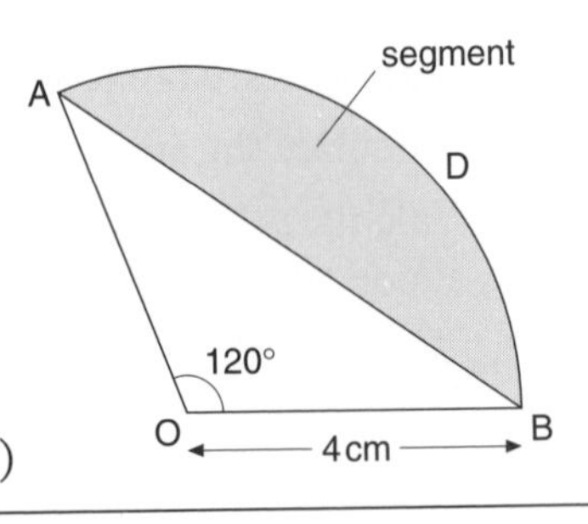

To work out the area of the segment ADB, work out the area of the sector OADB and subtract the area of the triangle OAB.

Area of sector = $\frac{120}{360} \times \pi \times 4^2\text{cm}^2 = 16{\cdot}755\ldots\text{cm}^2$

Area of triangle OAB = $\frac{1}{2} \times \text{OA} \times \text{OB} \times \sin 120° = 6{\cdot}928\ldots\text{cm}^2$

So area of segment = $(16{\cdot}755\ldots - 6{\cdot}928\ldots)\text{cm}^2 = 9{\cdot}826\ldots\text{cm}^2$
$= 9{\cdot}83\text{cm}^2$ (to 3 s.f.)

1 A conical firework has a base radius of 8·2 cm and a height of 14·3 cm.

(a) Calculate the volume of the firework.

(b) The label on the firework covers the entire sloping face. Calculate the surface area of the label.

WJEC

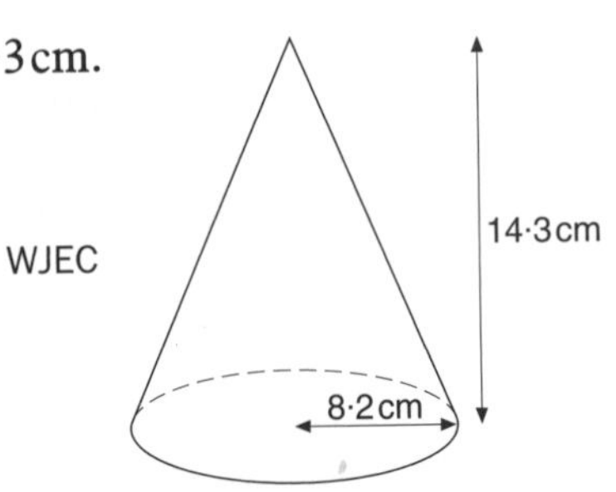

2 The game of boules is played with hollow steel balls. Each ball has an outside diameter of 80 mm, and the thickness of the steel is 5 mm.

Calculate the volume of steel required to make one ball.

3 An ice cream cone has height 11 cm and radius 3·5 cm.
Ice cream completely fills the cone and forms a hemisphere on the top of the cone.

Neglecting the thickness of the cone, calculate the volume of ice cream.
Give your answer to an appropriate degree of accuracy.

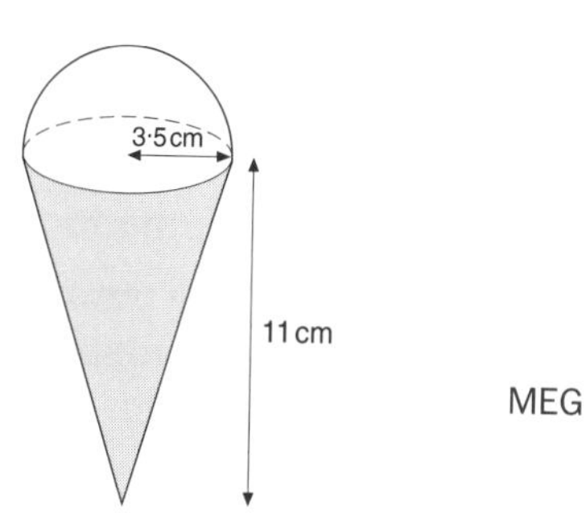

MEG

4 This glass test tube is cylindrical with a hemispherical base.
The internal diameter is 2·6 cm.
It is filled with water to a depth of 11·3 cm.

2·6cm
11·3cm

(a) Calculate
  (i) the volume of water in the test tube,
  (ii) the area of glass in contact with the water.

(b) A stone sinks into the water in the test tube, causing the water level to rise by 24 mm. Calculate the volume of the stone.

NICCEA

5 Longdon goatsmilk cheese is sold in small portions.
The cross section of each portion is a sector of a circle, radius 5 cm.
The angle at the centre is 40° and the portion is 1·5 cm thick.

(a) What is the volume of each portion?

(b) The top of each portion carries a triangular label, as shown.
What is the area of the top of the cheese **not** covered by the label?

6 The diagram shows glass sectors in a wooden door.
The angles of the sectors are 40°, 60° and 40°,
and the radius of each sector is 24 cm.

(a) Calculate the total area of glass used in the sectors.

(b) Thin strips of wooden framing (which can be curved) are to be placed along the perimeter of each sector.
Calculate the total length of framing required, giving the answer in metres correct to 1 decimal place.

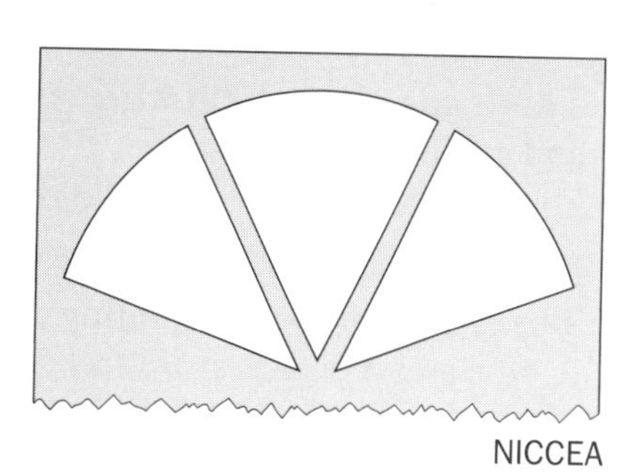

NICCEA

7

3·5cm
A B
cup
2·1cm
E C
15cm
stem
D
base

A wine glass is shown in the diagram.
The curved surface of the glass cup is part of a cone whose vertex D is at the base of the stem.
The radius AB of the top of the cup is 3·5 cm
and the radius of the bottom of the cup is 2·1 cm.
The overall height of the glass is 15 cm.

(a) Calculate the height DE of the stem.

(b) Calculate the volume of wine that the glass can hold.

WJEC

Answers and hints ► page 128

Transformations

Describing transformations

When you describe a transformation, you need to give enough information to specify it completely.
These are all transformations of A_1 to A_2.

This is a **translation** of **3 across** and **1 up**.
You must give the *across* and *up* distances.

You could give these as a **column vector** $\begin{bmatrix} 3 \\ 1 \end{bmatrix}$.

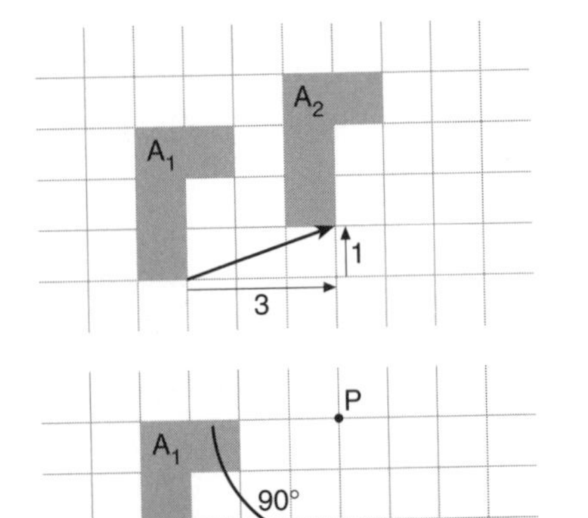

This is a **rotation** of **90°** anticlockwise, **centre P**.
You need to give the *angle* (and its direction) and the *centre*.

It is easiest to find the centre by trying out different points using tracing paper.

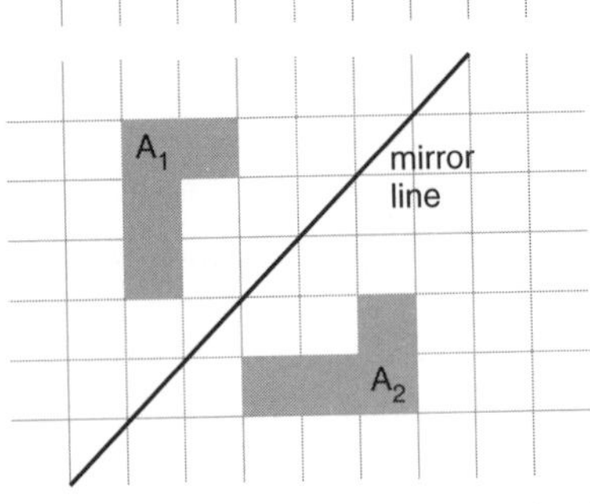

This is a **reflection** in the mirror line shown.
You only need to specify the *line* exactly.

You may have to draw the mirror line or give the equation of the line.
Use tracing paper to check your answer.

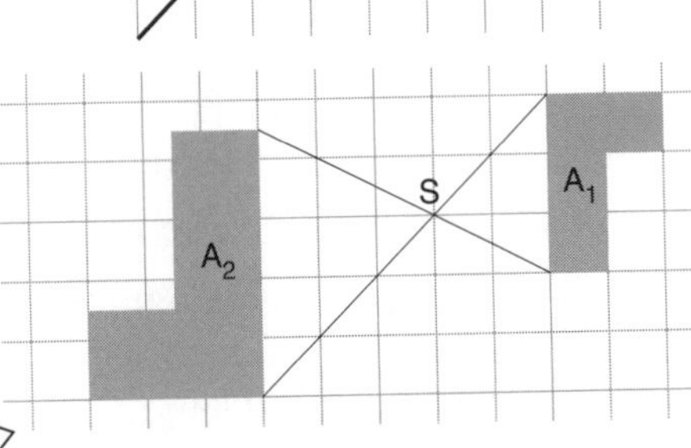

This is an **enlargement** with **centre S** and **scale factor** $^-$**1·5**.
You need to give the *centre* and the *scale factor*.

Note that the ratio of the sides of A_1 to A_2 is 1 to 1·5.
The ratio of the corresponding distances from A_1 to S and A_2 to S is also 1 to 1·5.

When a shape is enlarged (or reduced), we say that the new shape is **similar** to the first one.

Scales and similarity ► page 60

Strip patterns

One way to make repeating strip patterns is by repeated rotations.

T_1 is rotated 180° about A.

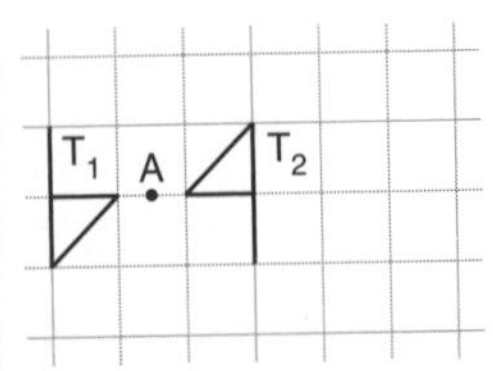

Both T_1 and T_2 are rotated about a different centre, B.

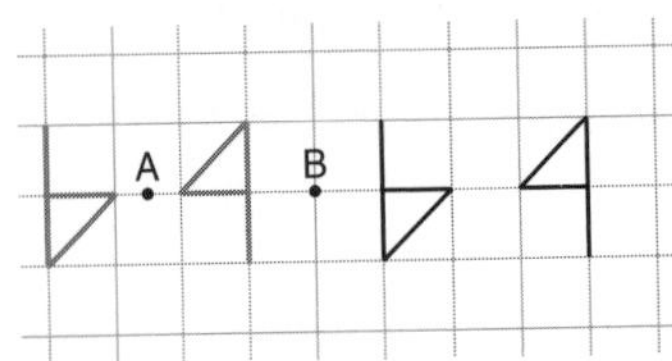

The whole pattern is rotated again about the first centre, A. Then rotate again about B and repeat for ever!

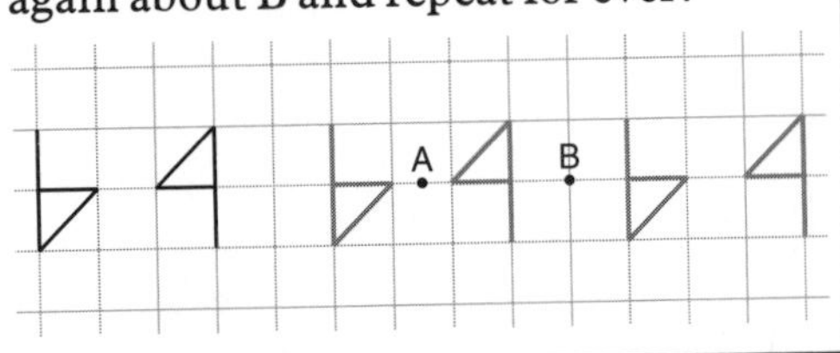

1 The diagram shows an arrangement of six square floor tiles.
Describe completely the single transformation which maps

(a) the tile CDGH onto the tile CBIH,

(b) the tile CDGH onto the tile GDEF.

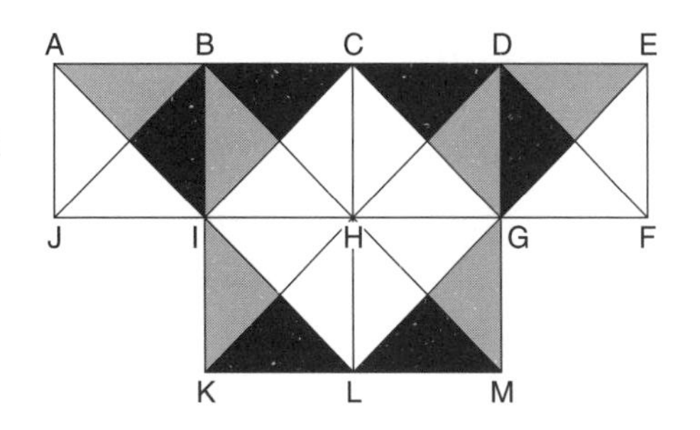

MEG

2 Paul is stencilling a border along his bedroom wall.
He wants the border to be a repeating pattern with 2-fold symmetry and a period of 5 cm.
He has begun using the stencil on the right.
Copy the diagram onto squared paper and show as much as you can of Paul's border.

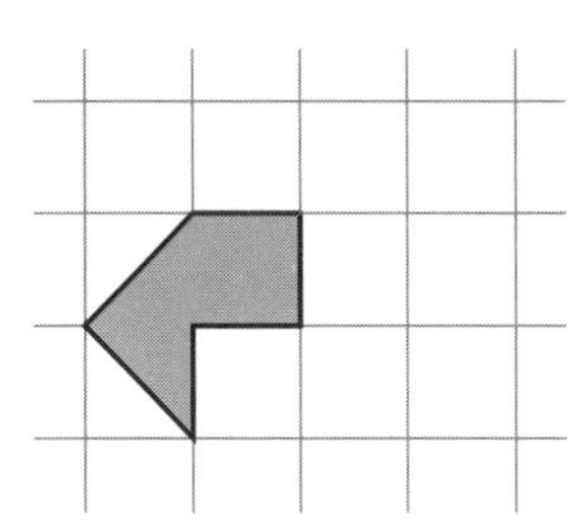

3

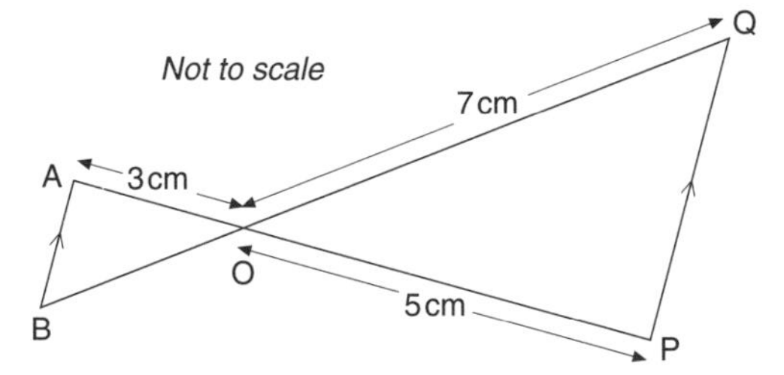

In the diagram, BA is parallel to PQ.
Triangle OAB is mapped onto triangle OPQ by means of a **single** transformation.

(a) Give a complete description of this transformation.

(b) Calculate the length of OB.

MEG

4 (a) John maps square P on to square Q by a translation.
Write the column vector for this translation.

(b) Sylvia maps square P on to square Q by a half-turn rotation.
Write down the coordinates of the centre of rotation.

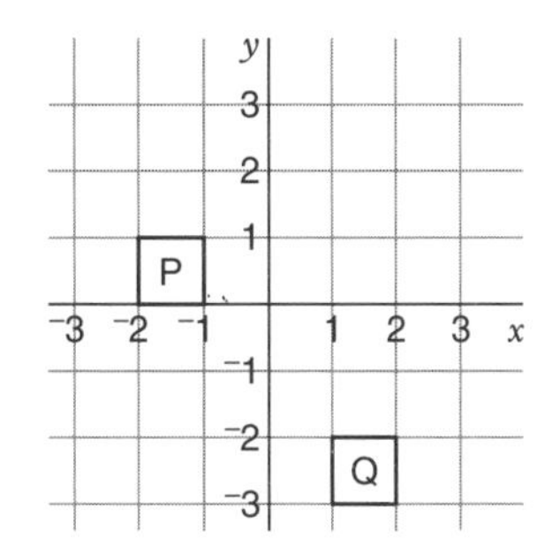

MEG/ULEAC (SMP)

5 On squared paper label an x-axis from ⁻6 to 6 and a y-axis from ⁻4 to 7.
Draw the triangle with coordinates A (1, 1), B (2, 1) and C (1, 3).

(a) The triangle ABC is enlarged with centre O and scale factor ⁻2 to form triangle A'B'C'. Draw and label A'B'C'.

(b) The original triangle ABC is then rotated about the point (1·5, 0) through an angle of 180° to form A"B"C". Draw and label A"B"C".

(c) What single transformation will take A'B'C' to A"B"C"?
Describe it completely.

Question 6 is on worksheet H5.

Answers and hints ► page 129

Scales and similarity

Scales and areas on maps

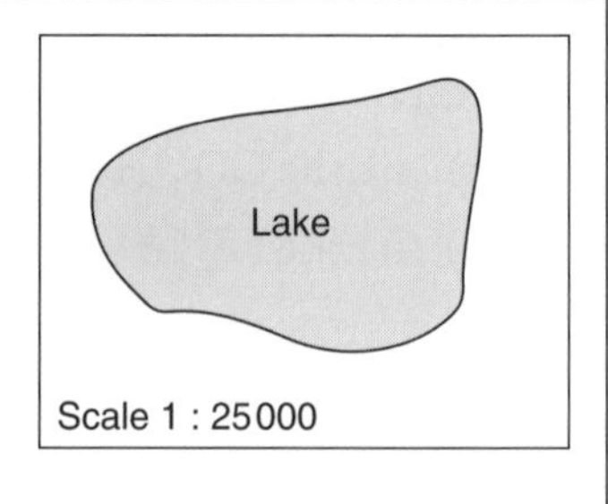

On the map, the lake has area $4\,\text{cm}^2$. What is its real area in km^2?

1 cm on the map is 25 000 cm on the ground.
25 000 cm = 250 m = 0·25 km.

So $1\,\text{cm}^2$ on the map is $0{\cdot}25 \times 0{\cdot}25\,\text{km}^2$ on the map $= 0{\cdot}0625\,\text{km}^2$.

Therefore $4\,\text{cm}^2$ on the map is really $4 \times 0{\cdot}0625\,\text{km}^2 = 0{\cdot}25\,\text{km}^2$.

Similar objects

Example

These glasses are similar.
The area of the base of the small glass is $30\,\text{cm}^2$ and it holds $400\,\text{cm}^3$.
Find h and the base area and capacity of the large glass.

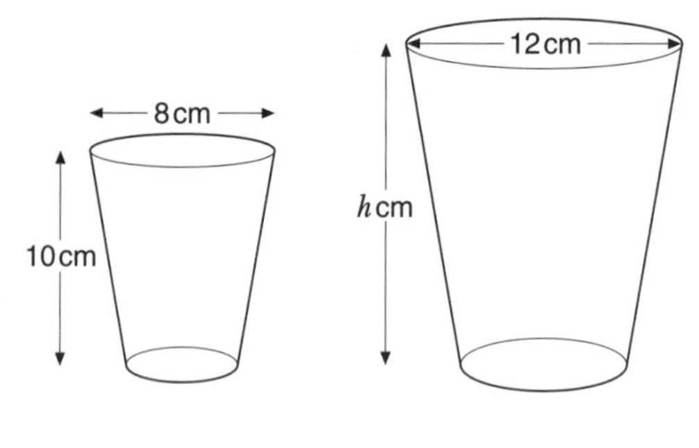

The glasses are similar, so the large glass is an enlargement of the smaller. Looking at the tops, we can see that the scale factor is $\frac{12}{8} = 1{\cdot}5$.
So the length h cm $= 1{\cdot}5 \times 10\,\text{cm} = 15\,\text{cm}$.

If the length scale factor is 1·5, then the area scale factor is $1{\cdot}5^2$.
So the base area of the large glass is $30 \times 1{\cdot}5^2\,\text{cm}^2 = 67{\cdot}5\,\text{cm}^2$.

If the length scale factor is 1·5, then the volume scale factor is $1{\cdot}5^3$.
So the volume of the large glass is $400 \times 1{\cdot}5^3\,\text{cm}^3 = 1350\,\text{cm}^3$.

Enlargement ► page 58
Ratio and proportion ► page 10

Remember:
Area scale factor is the **square** of the length scale factor.
Volume scale factor is the **cube** of the length scale factor.

1 The scale on an Ordnance Survey map is 1: 50000.

(a) On the map, the distance from Longdon to Upton is 12 cm. What is the actual distance in kilometres?

(b) Upton Meadows have an area of $20\,\text{cm}^2$ on the map. What is the real area of the Meadows in square kilometres?

2 The Popigai Crater in Siberia is a large circular crater made by the impact of an asteroid or comet.
The diameter of the crater is 100 km.

A map of Siberia is drawn to a scale of 1 to 8000000.
What is the diameter of the crater on the map in millimetres?

MEG/ULEAC (SMP)

3 On a clear day, if you stand at Nefyn in North Wales (sea level) and look towards Snowdon, another mountain, Garnedd-goch, blocks the view. From the map, the distance from Nefyn to Garnedd-goch is 21 km, and from Garnedd-goch to Snowdon is 11 km.

The height of Snowdon is given as 1085 m.
What is the lowest the height of Garnedd-goch could be?

MEG/ULEAC (SMP)

4 A map of part of the Yorkshire Dales is drawn to a scale of 1:50000.

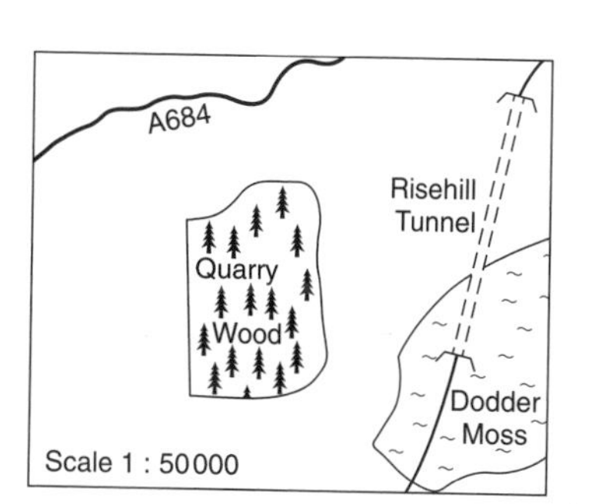

(a) On the map the length of Risehill Tunnel is 2·3 cm. Calculate the actual length of the tunnel in kilometres.

(b) Quarry Wood covers an area of 2 cm^2 on the map. Calculate the actual area of Quarry Wood in hectares. (1 hectare = 10000 m^2)

MEG

5 In Xian, China, you can buy solid scale models of the famous Terracotta Warriors. A model 16 cm tall weighs 270 grams and has an armour plate of area 9 cm^2.

(a) Calculate the armour plate area on a similar model of height 24 cm.

(b) Calculate the weight of the same 24 cm model.

MEG (SMP)

6 A £1 coin has diameter 22 mm.
In 1989 a £2 coin was issued to commemorate the tercentenary of the Bill of Rights. Its diameter is 28 mm. Its mass is double that of the £1 coin and it is made of the same metal.

Are they mathematically similar solids (apart from the designs)?
Give reasons and show your working.

Reproduced by permission of the Royal Mint

MEG/ULEAC (SMP)

7 Sudsy shampoo comes in plastic bottles.
Two different sizes are available, and the two bottles are similar.
The height of the large size is 1·3 times the height of the small size.
The small size costs £1·40 and the large size costs £2·40.

Which size gives you more for your money?
Explain clearly how you worked it out.

8

At dinner in a restaurant in France, we had on our table a 50 cl jug for wine and a 25 cl jug for milk.
The jugs were mathematically similar in shape.
The smaller jug was 15 cm high.

What was the height of the larger jug?

MEG/ULEAC (SMP)

Answers and hints ► page 130

Units and measures

Converting beween 'old' and metric units and vice versa

You need to know all these approximations.

1 pound (lb = 16 ounces) is about $\frac{1}{2}$ kg.
1 inch (in or ") is about 2·5 cm or 25 mm.
1 foot (ft or ' = 12 inches) is about 30 cm.
1 mile is about 1·6 km.
1 pint (pt) is about $\frac{1}{2}$ litre.
1 gallon (= 8 pints) is about 4·5 litres.

1 metre is about 40 inches.
1 kilometre is about $\frac{5}{8}$ mile.
1 kilogram is just over 2 pounds. (2·2 lb)
1 litre is just under 2 pints. ($1\frac{3}{4}$ pints)

Converting metric units

1 centimetre (cm) = 10 millimetres (mm)
1 metre (m) = 100 cm = 1000 mm
1 kilometre (km) = 1000 m
1 kilogram (kg) = 1000 grams (g)

1 litre = 100 centilitres (cl)
= 1000 millilitres (ml) ⇐ 1 ml = 1 cm^3
1 cubic metre (m^3) = 1000 litres
1 tonne = 1000 kg

Rates

Speed If I travel 80 km in 5 hours, my average speed is 80 km/5 h = 16 km/h.

Density If a block of metal with volume 450 ml has a density of 7·6 g/ml, then its mass is $450 \times 7{\cdot}6$ g = 3420 g = 3·42 kg.

Upper and lower bounds

Rounding ► page 4

If you measure 'to the nearest cm', then the measurement may be out by 0·5 cm.
'The length is 4 cm to the nearest cm' means the real length is between (4 − 0·5) cm and (4 + 0·5) cm; that is between 3·5 cm and 4·5 cm.

'The weight is 4 kg to the nearest 100 g' means the real weight is between

$$(4\text{kg} - 0{\cdot}5 \times 100\text{g}) \quad \text{and} \quad (4\text{ kg} + 0{\cdot}5 \times 100\text{g})$$

or $4000\text{g} - 50\text{g} \le \text{real weight} < 4000\text{g} + 50\text{g}$

Lower bound ⇒ $3950\text{ g} \le \text{real weight} < 4050\text{g}$. ⇐ Upper bound

Example
A car travels 120 m in 5·8 seconds. If both figures are correct to 2 s.f., what is the fastest average speed at which the car could have been travelling?

Car's speed = distance ÷ time.
The longest distance it could have travelled is 125 m and the shortest time is 5·75 s.
With these figures, the fastest speed = 125 ÷ 5·75 = 21·739... m/s = 21·7 m/s (to 3 s.f.).

1 Ela drove her car from Amsterdam to Apeldoorn in Holland, a distance of 137 km.
The appropriate accuracy for this distance is to the nearest kilometre.
Give the appropriate accuracies for the measurements of these quantities:

(a) the mass of a egg
(b) how long a television programme lasts
(c) the height of a tree
(d) the thickness of your skin.

MEG/ULEAC (SMP)

2 Arrange each of these in order of size, smallest first.
Show your working.

(a) 1 gallon, 750 centilitres, 3 litres
(b) 8 ounces, $\frac{1}{2}$ kilogram, 2 pounds
(c) 1 inch, 50 millimetres, 0·01 metres
(d) 1400 metres, 1 mile, 2 kilometres

3 Old 12-inch diameter records used to rotate at 78 revolutions per minute.
Roughly what was the speed of the outside edge of the record in centimetres per second?

4 An African fruit bat when trying to attract a mate can make an estimated 26 000 calls in an 8-hour day.
How many calls per minute is this?

MEG/ULEAC (SMP)

5 5p coins are made so that each coin has a mass of 3·26 g, to 2 decimal places.

(a) What is the smallest the mass of a 5p coin could be?

(b) What is the greatest mass a £5 bag of 5p coins could have?

(c) Is this sufficient error to invalidate the method of checking the number of coins in a £5 bag by weighing it?
Explain your reasoning.

Reproduced by permission of the Royal Mint

MEG/ULEAC (SMP)

6 The radius of a circle is measured as 4·7 cm.
Calculate the least possible value of its area.
(Take the value of π from your calculator or use 3·141 59.)

MEG (SMP)

7 A carton of apple juice has the shape of a cuboid with a square base of side 6·8 cm and a height of 18·4 cm.
Each of these measurements is correct to one decimal place.

(a) Given that 5000 such cartons must be filled, calculate the volume of juice necessary to be certain of filling all the cartons.

(b) Calculate the maximum number of cartons that could be filled with this volume of juice.

MEG

8

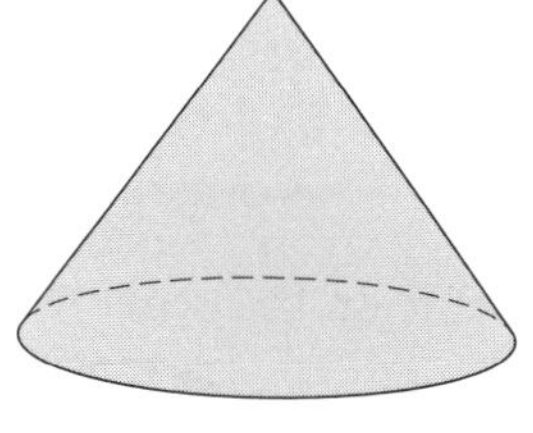

Chris wants to know how much coal there is in a coal heap.
The heap is in the shape of a cone. He estimates that the height of the heap to the nearest 5 metres is 30 metres, and the diameter to the nearest 10 metres is 80 metres.
Use the formula $V = \frac{1}{3}\pi r^2 h$ to find upper and lower estimates for the volume of coal in the heap, showing the values of r and h that you are using.

MEG (SMP)

Answers and hints ► page 131

Loci

A locus is a set of points which fit a particular rule.
For example:

An ant moves so that it is always 2 cm from the point O.

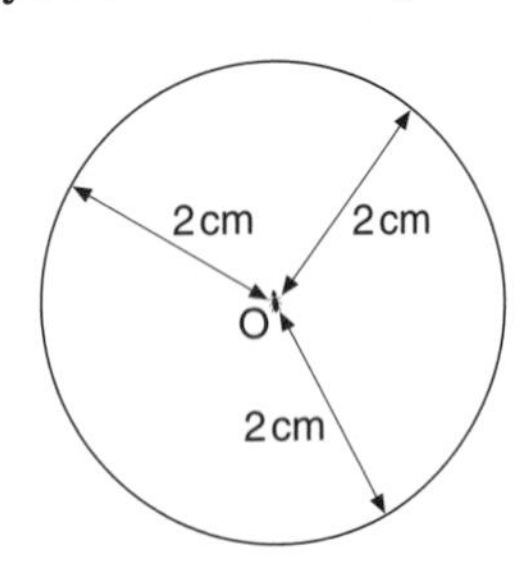

A dog can move so that it is always 5 m or less from the wall of a house.

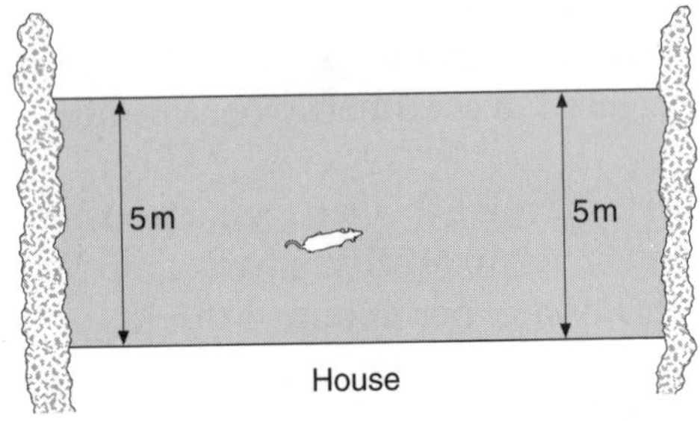

A ship sails so that it is the same distance from lighthouse A as it is from lighthouse B.

The donkey can eat grass inside and up to 50 cm outside the electric fence.

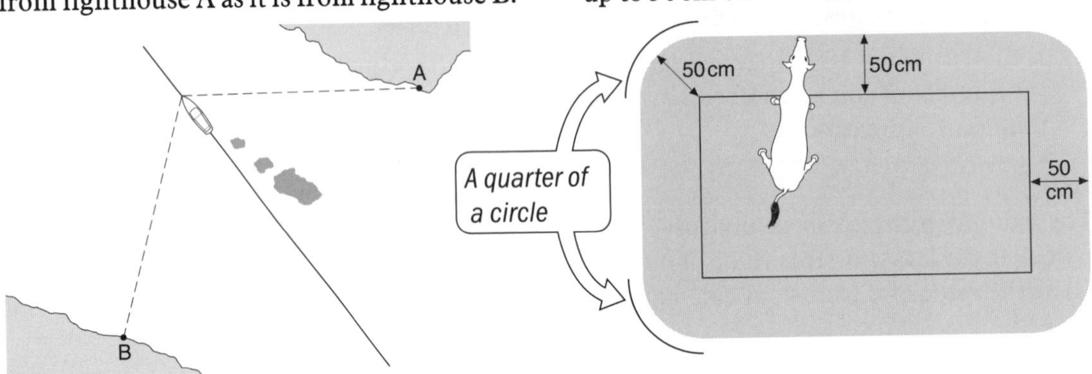

Constructing the locus of a point which is the same distance from both A and B.

The line is called the **perpendicular bisector of AB**.

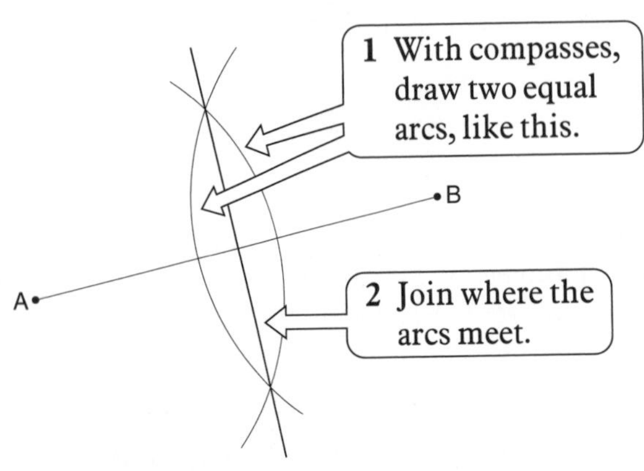

Constructing the locus of a point which is the same distance from the lines OA and OB.

The line is called the **bisector of angle AOB**.

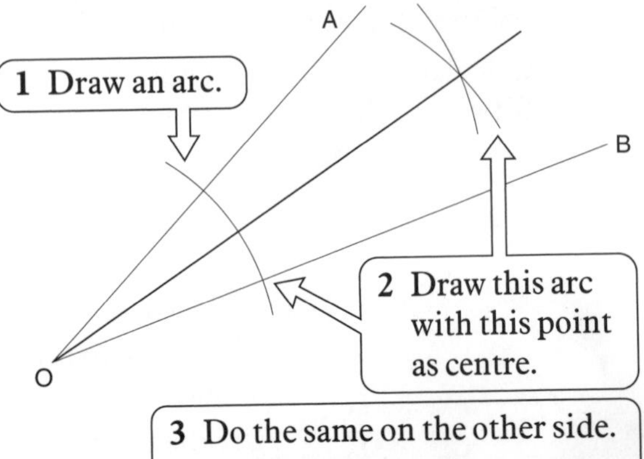

All the questions on loci are on worksheets H6 to H9.

Answers and hints ► page 132

Trigonometric functions

Graphs of $\sin x$, $\cos x$ and $\tan x$

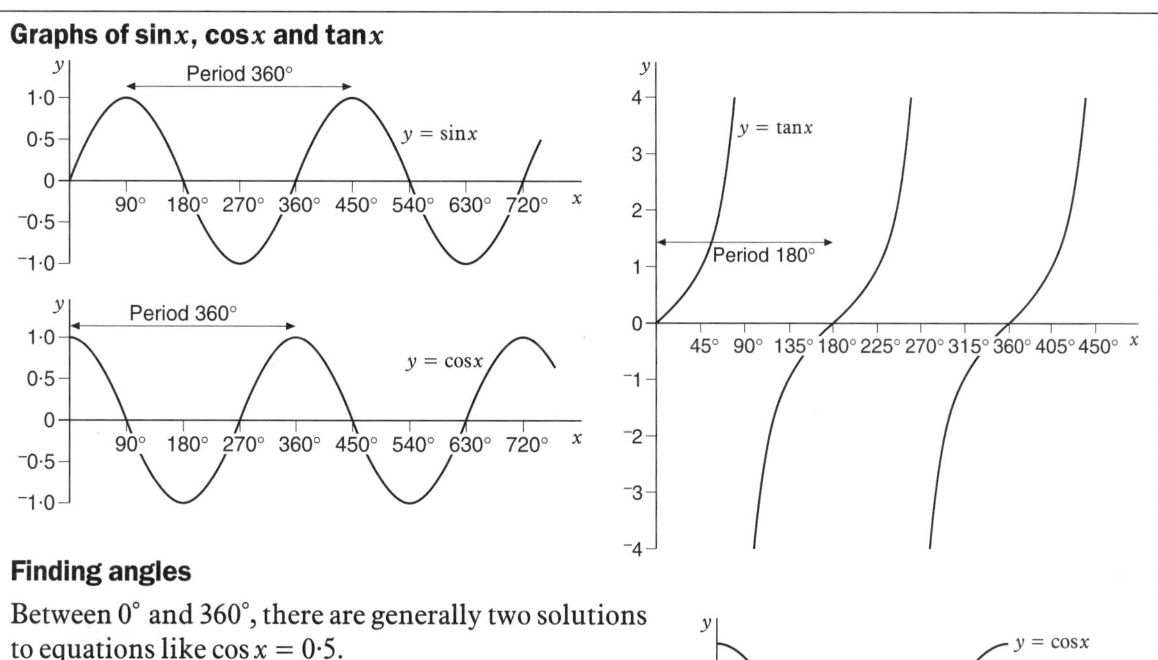

Finding angles

Between 0° and 360°, there are generally two solutions to equations like $\cos x = 0{\cdot}5$.
Find the first with your calculator:

[0] [·] [5] [$\cos^{-1}$] [=] [60] ⇐ *Check how to do this in your manual.*

Then use a sketch graph of the trigonometric function to work out the other.

$y = \cos x$
$360° - 60° = 300°$

Finally, always check your answers with your calculator.
So here, check that $\cos 300° = 0{\cdot}5$.

Other trigonometric functions

Functions and graphs ► page 36

You may be asked to sketch graphs like $y = 3\sin 2x$.

The graph is 3 times the height of $\sin x$. We say its **amplitude** is 3.

The graph repeats when $2x = 360°$, that is when $x = 180°$. We say its **period** is 180°.

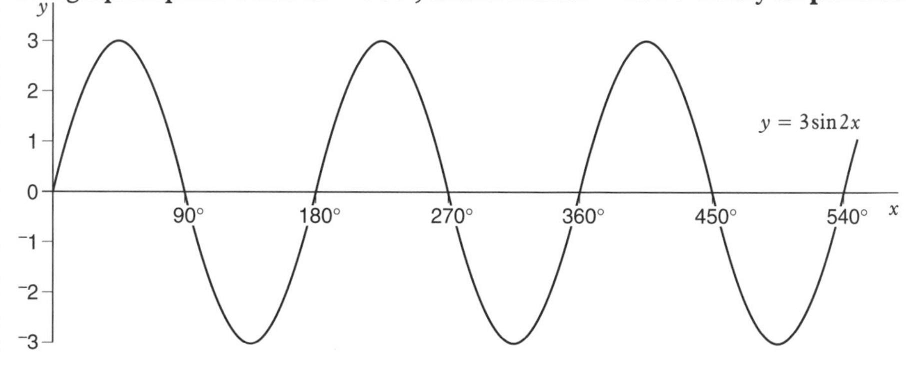

All the questions on trigonometric functions are on worksheets H10 to H13.

Answers and hints ► page 133

Vectors

Adding and subtracting vectors

If you have two vectors, $\underset{\sim}{a}$ and $\underset{\sim}{b}$, then...

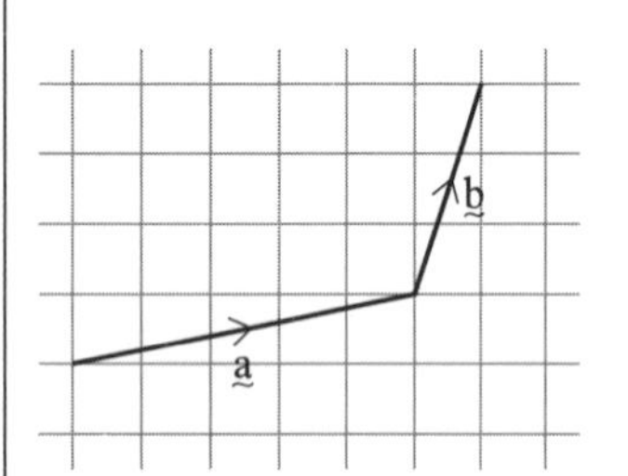

the **sum**, or **resultant**, $(\underset{\sim}{a} + \underset{\sim}{b})$ is found like this.

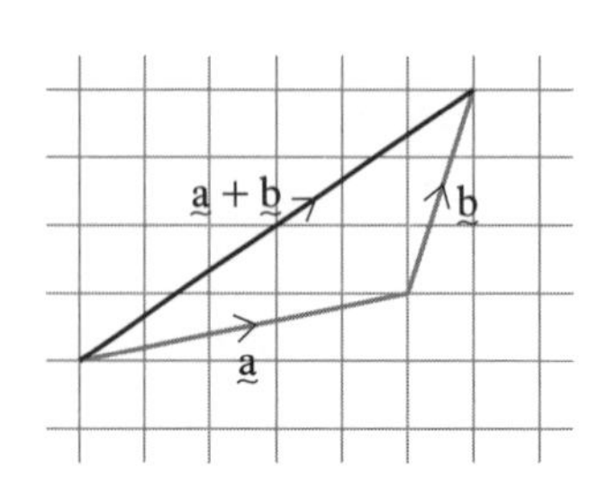

$\underset{\sim}{a} - \underset{\sim}{b}$ is found by reversing $\underset{\sim}{b}$ and adding, like this.

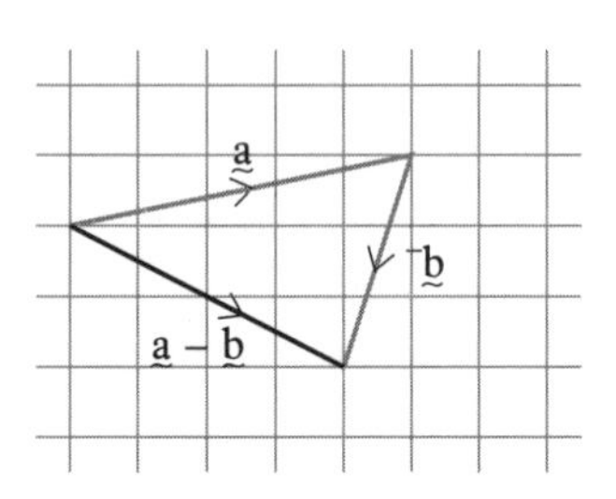

If $\underset{\sim}{a} = \begin{bmatrix} 5 \\ 1 \end{bmatrix}$ and $\underset{\sim}{b} = \begin{bmatrix} 1 \\ 3 \end{bmatrix}$, then...

$\underset{\sim}{a} + \underset{\sim}{b} = \begin{bmatrix} 6 \\ 4 \end{bmatrix}$

$\underset{\sim}{a} - \underset{\sim}{b} = \begin{bmatrix} 4 \\ ^{-}2 \end{bmatrix}$

Multiplication of vectors

You can multiply a vector by a number.

In this example, vector $\underset{\sim}{a} = \begin{bmatrix} 5 \\ 1 \end{bmatrix}$.

So $2\underset{\sim}{a} = 2 \times \begin{bmatrix} 5 \\ 1 \end{bmatrix} = \begin{bmatrix} 10 \\ 2 \end{bmatrix}$.

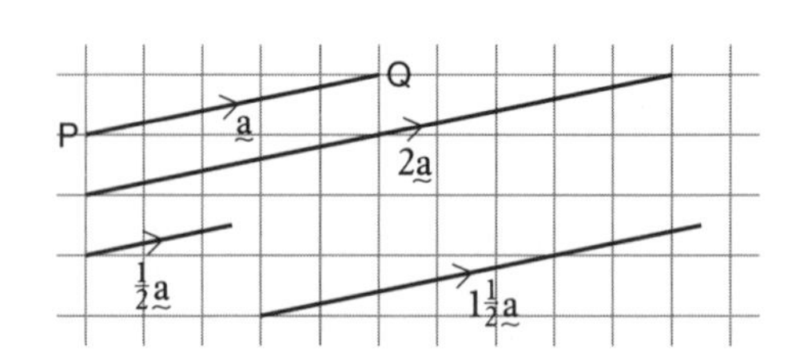

Vectors may be written in several ways.

The vector $\underset{\sim}{a}$ in the diagram above can be written $\underset{\sim}{a}$, **a** or $\overrightarrow{PQ}$.

When you write vectors by hand, write $\underset{\sim}{a}$ or $\overrightarrow{PQ}$.

1 David is taking part in an orienteering competition. The course ABCDE is shown on the plan.

A to B is $\begin{bmatrix} 3 \\ 1 \end{bmatrix}$.

(a) Write down B to C, C to D, D to E as column vectors.

(b) At D he makes a mistake and does $\begin{bmatrix} ^{-}2 \\ ^{-}1 \end{bmatrix}$.

After this, what is the column vector of his course to reach E?

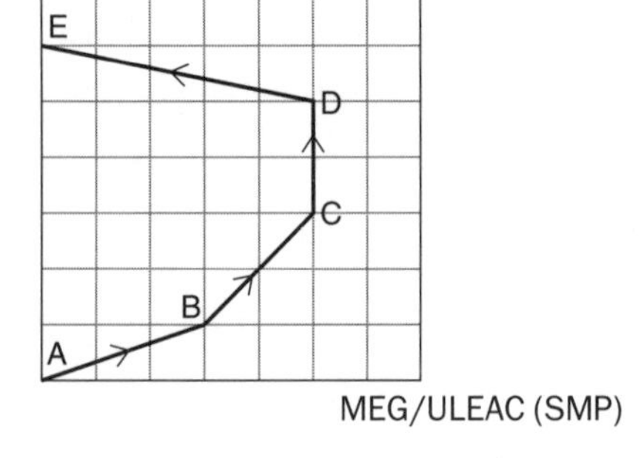

MEG/ULEAC (SMP)

2 A is the point $(^{-}1, 4)$, B is $(1, 1)$ and C is $(7, ^{-}8)$.

(a) Find the vectors

(i) $\overrightarrow{AB}$

(ii) $\overrightarrow{AC}$

(b) What can you deduce about the points A, B and C?

MEG (SMP)

3 In the diagram, the centre of a regular hexagon, ABCDEF, is at O.
The vectors $\overrightarrow{OA} = \underset{\sim}{a}$ and $\overrightarrow{OB} = \underset{\sim}{b}$.

(a) What is the special name given to the quadrilateral ABOF?

(b) Write down, in terms of $\underset{\sim}{a}$ and $\underset{\sim}{b}$, the vectors:

(i) $\overrightarrow{EF}$ (ii) $\overrightarrow{CD}$ (iii) $\overrightarrow{DA}$
(ii) $\overrightarrow{AB}$ (v) $\overrightarrow{EA}$

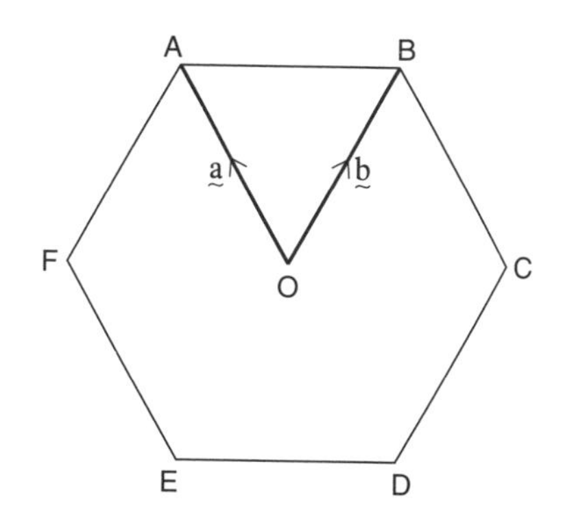

WJEC

4 In the diagram, $\overrightarrow{OA} = 2\underset{\sim}{a}$, $\overrightarrow{OB} = 3\underset{\sim}{a} - 2\underset{\sim}{b}$ and $\overrightarrow{OC} = 5\underset{\sim}{a} - 6\underset{\sim}{b}$.

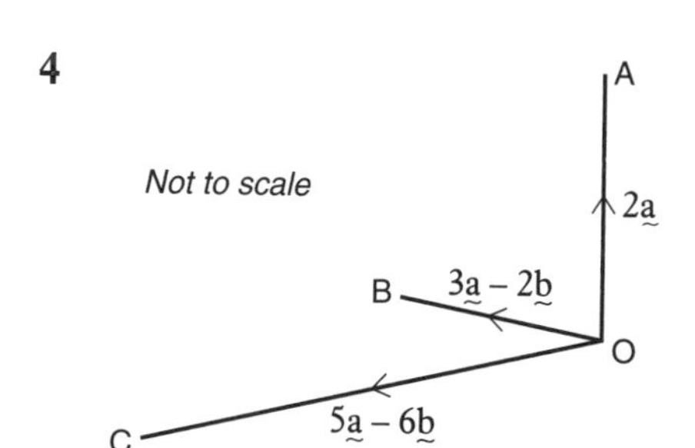

(a) Express, in terms of $\underset{\sim}{a}$ and $\underset{\sim}{b}$, as simply as possible,

(i) $\overrightarrow{AB}$

(ii) $\overrightarrow{BC}$

(b) What do your answers to part (a) tell you about the points A, B, and C?
Give a reason for your answer.

MEG

5 The diagram shows two forces pulling on an object.

(a) Draw a vector diagram to scale to show the addition of these two forces.

(b) Hence or otherwise find

(i) the magnitude of the resultant force on the object,

(ii) the angle that the resultant force makes with the direction of the 6 N force.

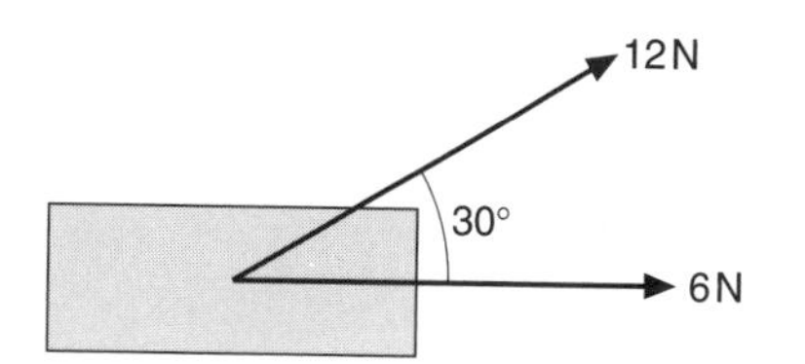

MEG/ULEAC (SMP)

6 Line AB has been enlarged by a scale factor of 2 with the centre of enlargement at O.
The enlarged line is A′B′.

$\overrightarrow{OA} = \underset{\sim}{a}$ $\overrightarrow{OB} = \underset{\sim}{b}$

(a) Express the following in terms of $\underset{\sim}{a}$ and $\underset{\sim}{b}$:

(i) $\overrightarrow{OB'}$ (ii) $\overrightarrow{AA'}$ (iii) $\overrightarrow{AB}$ (iv) $\overrightarrow{B'A'}$

A point C is chosen so that $\overrightarrow{OC} = \underset{\sim}{a} + \underset{\sim}{b}$.

(b) Find $\overrightarrow{CA'}$.

(c) Using your answers to (a) (iv) and (b), write down a relationship between $\overrightarrow{B'A'}$ and $\overrightarrow{CA'}$.

(d) What does this tell you about the points B′, C and A′?

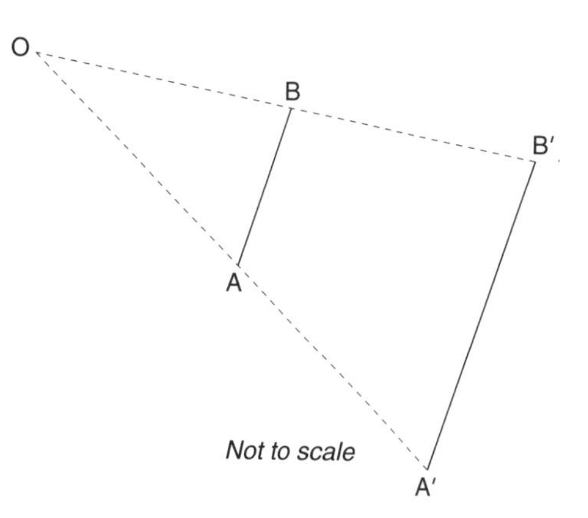

MEG/ULEAC (SMP)

Answers and hints ► page 135

Mixed shape, space and measures

None of the diagrams is drawn to scale.

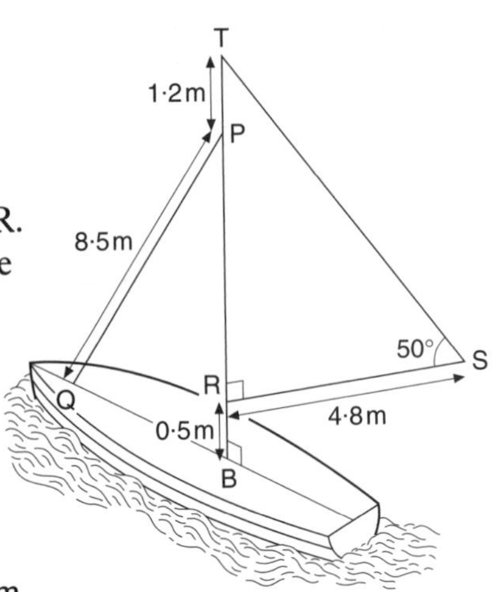

1 The triangular sail, TRS, of a small boat has a right-angle at R. The sail is hinged to a vertical mast, TRB, so that it can rotate with RS remaining horizontal.

(a) Describe the locus of S relative to the boat as the sail moves.

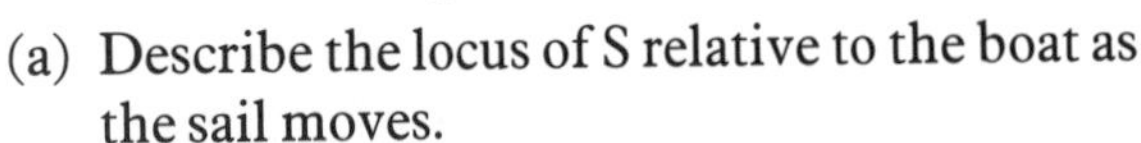

(b) RS = 4·8 m, angle TSR = 50° and RB = 0·5 m.
Calculate the height, TB, of the mast.

(c) P is a point on the mast 1·2 m below T.
The mast is held in position by a rope, PQ, of length 8·5 m.
Calculate the acute angle between PQ and the mast.

MEG

2 The great Pyramid of Cheops in Egypt, which for the purposes of this question should be regarded as an exact solid pyramid, has a horizontal square base of side 230 metres and a vertical height of 146·5 metres.
The vertex is vertically above the centre of the base.

(a) One cubic metre of the rock used to build the pyramid has a mass of 2·80 tonnes.
Calculate the mass, in tonnes, of the pyramid.
Give your answer in standard form, correct to three significant figures.

(b) Calculate the total area, in square metres, of the four slant faces of the pyramid.
Give your answer correct to three significant figures.

MEG

3 The diagram shows a circle, centre O, with points A, B, P and Q on the circumference. BOQ is a diameter.

$\overrightarrow{OA} = 2\mathbf{a}$ $\quad\overrightarrow{PQ} = 3\mathbf{a}$ $\quad\overrightarrow{OB} = 2\mathbf{b}$

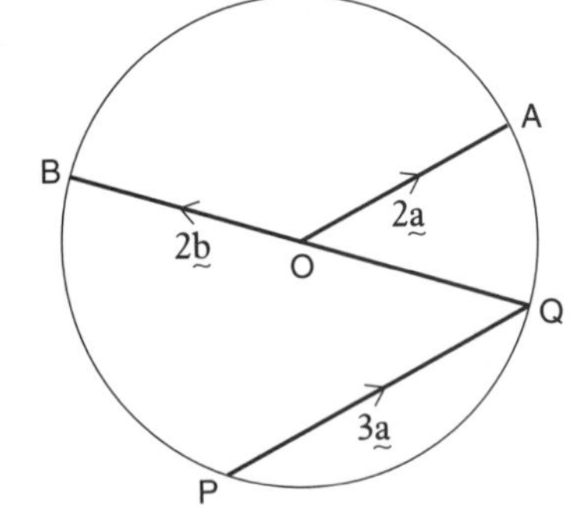

(a) Write down, in terms of $\mathbf{a}$ and $\mathbf{b}$, the vectors

(i) $\overrightarrow{AB}$ (ii) $\overrightarrow{QB}$ (iii) $\overrightarrow{PB}$

(b) If $\mathbf{a} = \begin{bmatrix} 4 \\ 3 \end{bmatrix}$, show that the length of vector $\mathbf{a}$ is 5 units.

(c) Use the result in part (b) to write down the lengths of
(i) OA (ii) BQ (iii) PQ

(d) Explain why angle BPQ is 90°.

(e) Calculate
(i) the length of BP
(ii) the size of angle BQP
(iii) the size of angle BAO

MEG

4 (a) Draw accurately a regular pentagon with its vertices on the circumference of a circle of radius 6 cm.

(b) Use measurements from your drawing to calculate the area of the part of the circle which is not inside the pentagon.

MEG/ULEAC (SMP)

5 A surveyor, standing on horizontal ground, is using a theodolite to measure the height of a vertical building.
She measures the distance from the foot of the building as 83·4 metres and the angle of elevation of the top as 32°.
The height of the theodolite is 1·6 metres.

(a) Calculate the height of the building in metres.

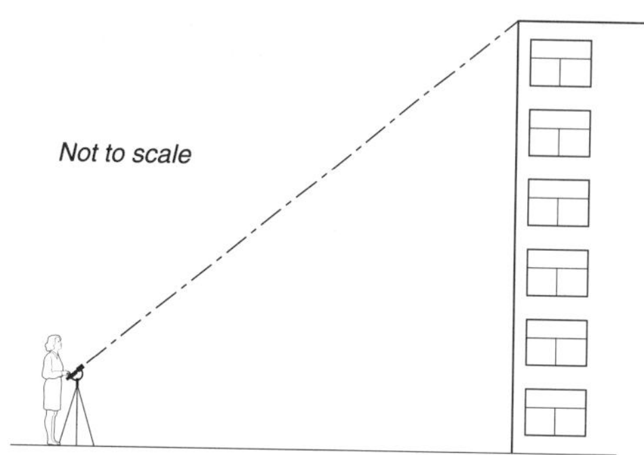

The measurement of the distance is accurate to the nearest 0·1 metre and the angle is accurate to the nearest degree.
The height of the theodolite is assumed to be exact.

(b) Calculate the greatest possible height of the building in metres.

MEG (SMP)

6 A bicycle chain is enclosed in a casing. The diagram shows the side of this casing.
The shapes ACB and RST are parts of circles. The marked lengths are in millimetres.

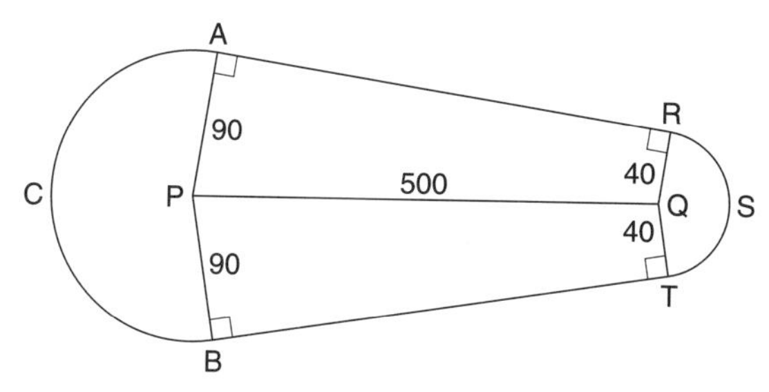

(a) Show that AR differs in length from PQ by approximately 2·5 mm.

(b) Find the area of ARQP.

(c) Show that the size of angle APQ is about 84·3°.

(d) Calculate the area of the side of the casing.

MEG/ULEAC (SMP)

7 The shape of a cut diamond is an octahedron, ABCDEF.
The plane BDEF is a square of side 0·8 mm.
Each of the slant edges AB, AD, AE, AF, CB, CD, CE and CF are 0·6 mm.
Calculate:

(a) the length of the diagonal BE,

(b) the angle between AB and the plane BDEF,

(c) the height of the diamond, AC,

(d) the volume of the diamond.

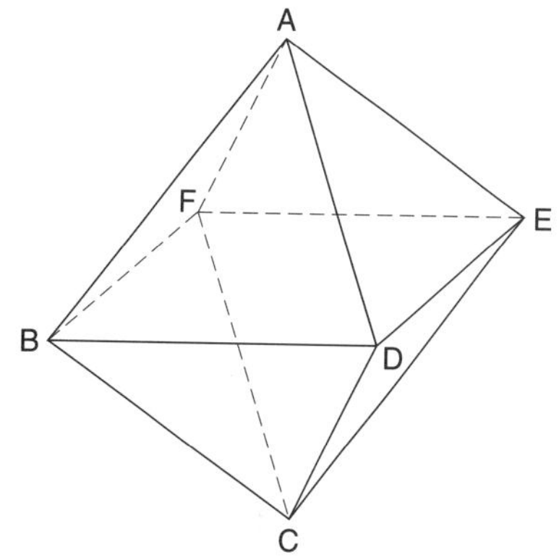

WJEC

8 Copy the axes and the three triangles ABC, LMN and PQN onto full size graph paper.

(a) On the same axes:

(i) Enlarge triangle ABC by a scale factor of 3, centre the origin.
Label the new triangle $A_1B_1C_1$.

(ii) Reflect triangle LMN in the y-axis.
Label the new triangle $L_1M_1N_1$.

(iii) Rotate triangle ABC 90° anticlockwise about (1, 0).
Label the new triangle $A_2B_2C_2$.

(b) Write down the vector representing the translation which maps triangle ABC onto triangle LMN.

(c) Which single transformation maps triangle LMN onto triangle PQN?

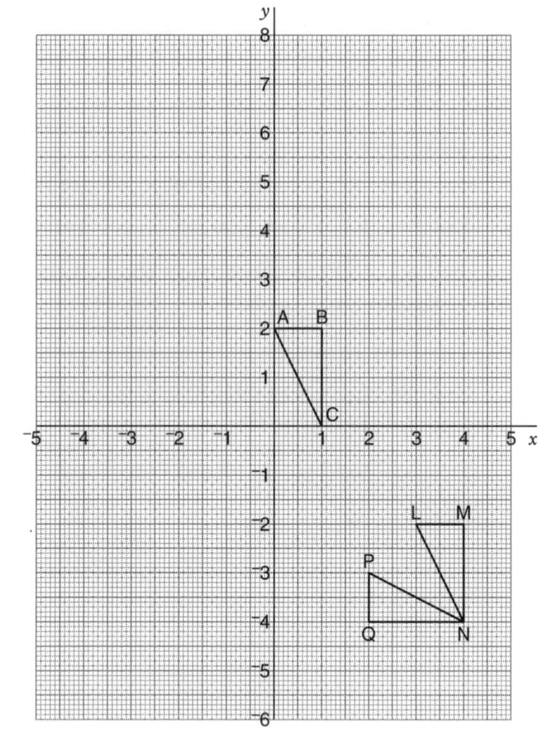

WJEC

9 **(In this question take π to be 3·14.)**

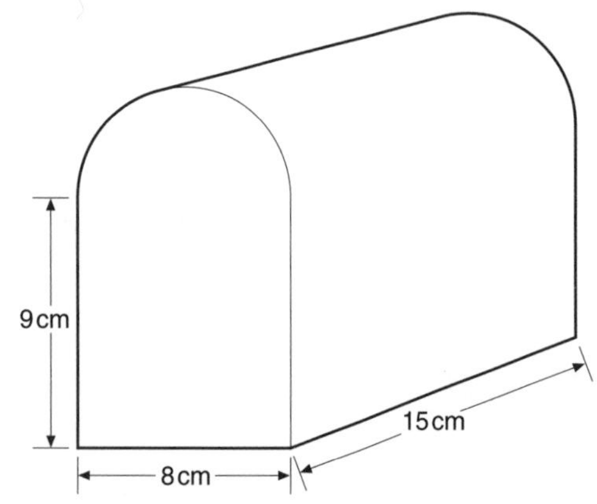

A loaf of bread has the shape of a prism, 15 cm long.
Its cross-section consists of a rectangle 8 cm by 9 cm with a semi-circular top.

(a) Calculate the volume of the loaf of bread.

(b) For publicity, the manufacturer of the bread decides to have a large balloon made in a similar shape to the loaf.
The enlargement is to have a scale factor of 50.

(i) Calculate the volume of the balloon, in cubic metres, when it is fully inflated.

(ii) Calculate the area of material, in square metres, needed to make the balloon. MEG

Answers and hints ► page 136

HANDLING DATA

Pie charts

The easiest way to show or measure percentages on a pie chart is with a pie chart scale, which is marked from 0 to 100%.

If you have to use a protractor marked in degrees you can use conversions like these.

Crimes involving property	64%
Other crimes	36%

To draw a sector for 64% of the total, work out 64% of 360°, which is $360 \times 0{\cdot}64 = 230°$ (to nearest 1°).

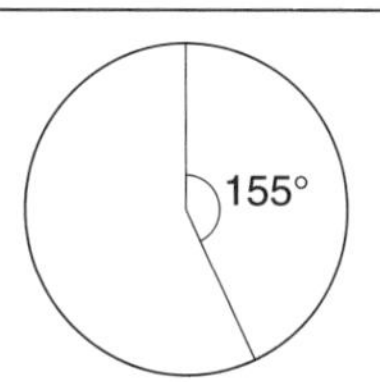

If you have measured an angle of 155°, the percentage it represents is $\frac{155}{360} \times 100 = 43°$ (to nearest 1°).

Percentages ► page 6

1 In 1985–86 the government spent its income as shown in the table:

	£ (billions)	%
Health and Social Security	56·5	42·8
Defence	18·1	
Education	14·2	
Housing and Environment	5·4	
Industry and Transport	8·5	
Other	29·3	
Total	132·0	

(a) Copy the table and complete the % column, giving your answer to one decimal place.

(b) Show the information in a pie chart. Label it clearly. MEG/ULEAC (SMP)

2 The pie chart shows the use of petroleum by different types of transport in Great Britain in 1990.

(a) What percentage of petroleum was used by cars and taxis?

(b) What percentage was used by ships and aircraft altogether?

(c) The total amount of petroleum used by transport in 1990 was 70 million tonnes. How many million tonnes were used by ships and aircraft altogether? Give your answer to the nearest million tonnes.

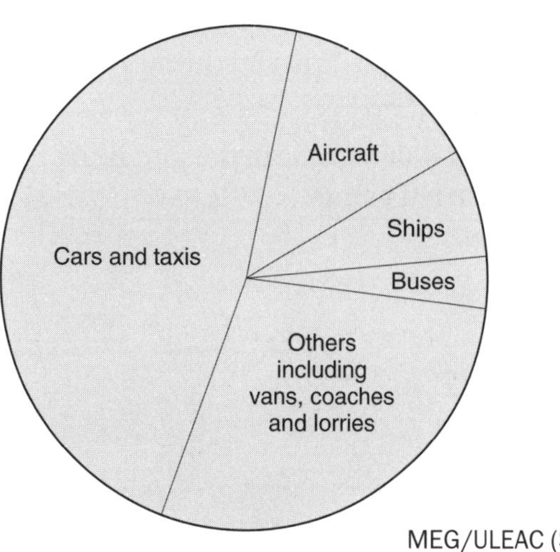

MEG/ULEAC (SMP)

Answers and hints ► page 139

Scatter diagrams

A scatter diagram is used to see whether there appears to be a relationship between two features, such as the handspan and height of the people in a particular group.

Each pair of values that relates to one member is plotted as a point.
Sometimes this gives a pattern where the points appear to lie clustered around a line – the 'line of best fit'. In such cases there is said to be **correlation** between the features.

These five examples are drawn to the same scale.

Strong positive correlation | **Some positive correlation**

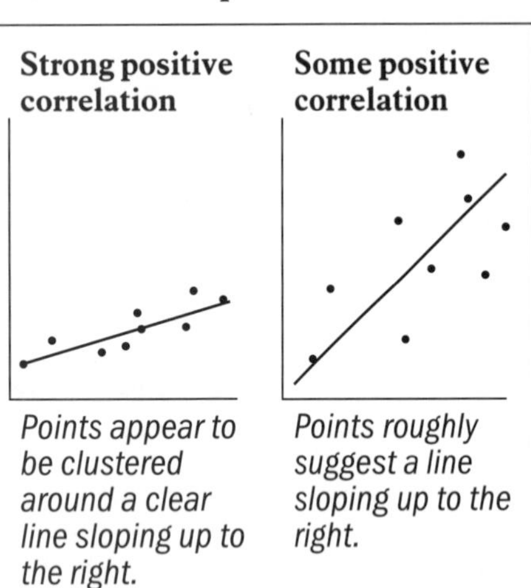

Points appear to be clustered around a clear line sloping up to the right.

Points roughly suggest a line sloping up to the right.

A positive correlation between the two sets of data means, roughly, 'the bigger the one, the bigger the other'.

Strong negative correlation | **Some negative correlation**

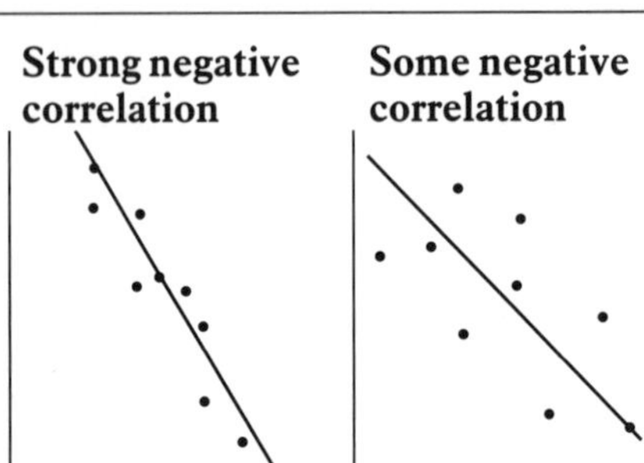

Points strongly suggest a line sloping down to the right.

Points appear to lie roughly on a line which slopes down to the right.

A negative correlation between the two sets of data means, roughly, 'the bigger the one, the smaller the other'.

No correlation

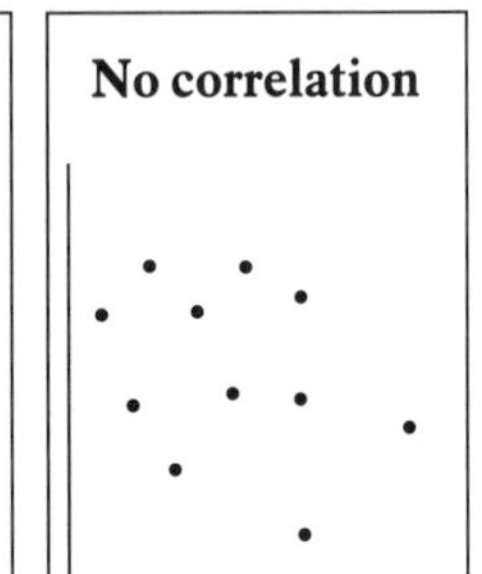

No obvious line emerges from looking at the points.

Even where there is correlation, you cannot say that one set of values is **determined** by the other set.

You may be asked to draw a line of best fit using the means of the values shown on the diagram.

Mean ► page 74

Suppose your diagram is about the dimensions of some aeroplanes.
The mean length is 25 m and the mean wingspan is 18 m.

Plot a point at (25, 18).
Check that it does come in the middle of your cluster of points.

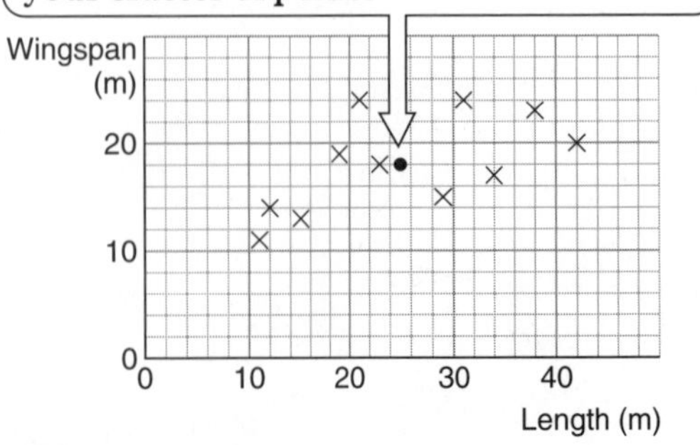

Now draw a line of best fit through it.

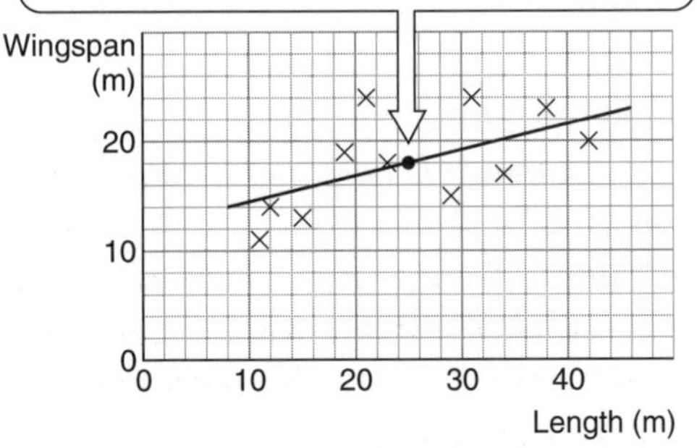

When you draw a line of best fit it helps to use a transparent ruler.

1 A group of pupils were tested on their multiplication tables.
They then did some intensive tables practice.
After the practice they were tested again.
This table shows their test scores before and after the test.

Score before practice	6	3	2	7	4	8	4	7
Score after practice	8	7	6	7	8	9	6	10

On graph paper, draw a scatter diagram for the data and draw a line of best fit.
Comment on how effective the practice was in improving their scores.

2 This table shows distances between pairs of cities, chosen at random from a map.
One distance is 'as the crow flies' measured on the map.
The other is the distance by a 'recommended' road route given in a mileage chart.

	Distance as the crow flies (miles)	Road distance (miles)
Nottingham to Lincoln	31	36
Birmingham to Leicester	34	42
Lincoln to Northampton	69	91
Nottingham to Oxford	82	103
Cambridge to Norwich	57	63
Oxford to Leicester	61	78
Colchester to Peterborough	67	89
Norwich to Lincoln	87	103
Coventry to Birmingham	17	19
Leicester to Coventry	22	26

(a) On graph paper, draw a scatter diagram for this information.

(b) Draw a line of best fit.

(c) Use what you have done to get a 'rule of thumb' for estimating the road distance if you know the distance between two cities 'as the crow flies'.

3 The table shows the scores of 12 pupils in maths and French tests.

Maths score	5	10	14	24	28	35	43	47	52	57	62	70
French score	12	6	12	24	18	24	33	23	29	35	28	35

(a) (i) Calculate the mean of the maths scores.
(ii) Calculate the mean of the French scores.

(b) (i) Plot the scores of these twelve pupils on a scatter diagram.
(ii) On the scatter diagram, plot the point whose coordinates are the mean scores you have calculated in part (a).
(iii) Draw the line of best fit on the scatter diagram.

(c) Vina scored 42 in the maths test but was absent for the French test.
Use your line of best fit to state a score she might have had for her French test if she had been present.

(d) Comment on this method.

MEG/ULEAC (SMP)

Answers and hints ► page 140

Mode, mean, median and range

Separate items of data

Suppose you have been given these fifteen items of data:

13 10 10 9 13 9 11 13 12 10 9 12 11 13 13

The **mode** is the value that occurs most frequently.

To find the **mean** add up the items and divide by the number of items.

Estimating the mean from grouped data ► page 80

Total = 168 Mean = $168 \div 15 = 11{\cdot}2$

If you put the items in order the one in the middle is the **median**.

If there are an even number of data items, there will be ***two*** *items in the middle. Add them up and divide by 2 to get the median.*

9 9 9 10 10 10 11 11 12 12 13 13 13 13 13

To find the **range**, look for the smallest and the largest value. Calculate the difference between them.

Range = $13 - 9 = 4$

Cumulative frequency ► page 76

Data in a table

Our fifteen items of data could be recorded in a table.

Value	9	10	11	12	13
Frequency	3	3	2	2	5

You can work out that the middle (eighth) item is in here, so 11 must be the median.

This shows that 13 is the mode.

To calculate the total of all the data items you can work out $(9 \times 3) + (10 \times 3) + \ldots$

Then you divide by the number of items to get the mean.

The number of items is the sum of the frequencies.

1 Leslie and Pat are members of a quiz team.
Here are their scores for the last eight quizzes.

Leslie	79	73	75	91	74	84	76	80
Pat	78	75	79	86	76	81	77	80

(a) Calculate Leslie's mean score.

(b) Calculate the range of Leslie's scores.

Pat has a mean score of 79 and a range of 11.

(c) If you were the captain of the quiz team, which of these players would you choose? Give a reason for your choice.

MEG/ULEAC (SMP)

2 A mathematics test was given to 30 boys. Their marks were as follows:

7	4	1	6	3	8	5	7	7	10
6	7	0	5	7	3	7	8	4	6
8	6	9	7	4	6	4	10	5	7

(a) Copy and complete this table.

Mark	0	1	2	3	4	5	6	7	8	9	10
Number of boys											

(b) Calculate the total of the marks obtained by the 30 boys.

(c) For the boys' marks, find
 (i) the mode,
 (ii) the median,
 (iii) the mean.

The mathematics test was also given to 20 girls.

(d) The girls' marks had a mean of 6·5.
 (i) Calculate the total of the marks obtained by the 20 girls.
 (ii) Calculate the mean of the marks obtained by the whole group of 50 boys and girls.

MEG

3 20 people from different families were asked how many newspapers and magazines were usually bought by the family per week.
The answers were:

Bill	3	Bob	10	Dave	14	Emma	8	Fran	10
Gemma	8	Jim	2	Kate	14	Ken	2	Mary	10
Mo	8	Pat	8	Paula	0	Pete	10	Phil	18
Sam	10	Sara	14	Steve	10	Sue	3	Tom	6

(a) What was the modal number of newspapers and magazines bought per week?

(b) What was the median number of newspapers and magazines bought per week?

(c) What was the mean number of newspapers and magazines bought per week?

NICCEA

4 Amal is the star scorer in her school's netball team.
Last term the mean number of goals she scored in a game was 11·2.
This term she wants to improve on her mean score.
She has played 8 games so far and her mean score is 10·5.

(a) How many goals has she scored so far this term?

(b) The next game will be the last game of the term.
How many goals at least must she score in it if she is to improve on last term's mean?

Answers and hints ► page 141

Cumulative frequency

This table is about the ages of the people in a village. There are 300 people altogether.

Age (x) in years	Frequency	Cumulative frequency
$0 \le x < 15$	56	56
$15 \le x < 30$	64	120
$30 \le x < 45$	65	185
$45 \le x < 60$	68	253
$60 \le x < 75$	47	300

This tells you there are 120 people less than 30 years old.

Check that you can tell how the other entries in this column were worked out.

You can use the values in the third column to plot a **cumulative frequency graph**. You can read off various statistical measures from the graph.

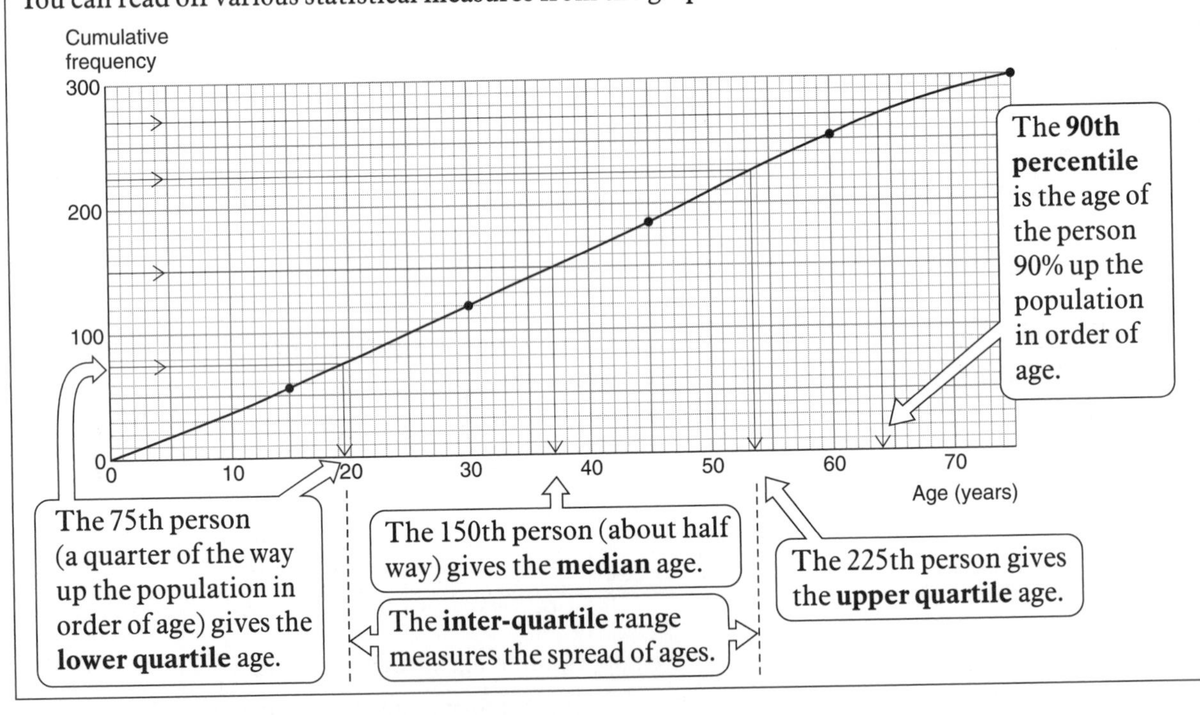

1 The table shows the distribution of the attendance figures at 40 football matches in England and Wales on a Saturday in January 1994.

Attendance 1000's	Frequency	Cumulative frequency
0–5000	15	15
5001–10000	12	27
10001–15000	2	
15001–20000	5	
20001–25000	2	
25001–30000	2	
30001–35000	2	

(a) Copy the table and complete the cumulative frequency column.

(b) Draw a cumulative frequency curve for this distribution.

(c) Use your cumulative frequency curve to estimate
  (i) the median attendance, (ii) the upper quartile.

MEG (SMP)

2 250 business telephone calls were made by a firm.
The lengths of these calls were recorded.
The results are shown in the table.

Length l (in minutes)	Number of calls	Cumulative frequency
$0 < l \leq 4$	24	
$4 < l \leq 8$	47	
$8 < l \leq 12$	68	
$12 < l \leq 16$	44	
$16 < l \leq 20$	32	
$20 < l \leq 24$	21	
$24 < l \leq 28$	8	
$28 < l \leq 32$	6	

(a) Copy the table and complete the cumulative frequency column.

(b) Draw the cumulative frequency curve.

(c) Find from the graph
  (i) the median length of the calls,
  (ii) the range of the hundred longest calls.

NICCEA

3 Jo is designing shorts for five-year-olds.
She needs to know how long to make the legs.
She measures the waist-to-knee length of 100 five-year-olds.
The results are in this table.

Waist-to-knee length w (cm)	Frequency
$32\text{ cm} < w \leq 34\text{ cm}$	3
$34\text{ cm} < w \leq 36\text{ cm}$	16
$36\text{ cm} < w \leq 38\text{ cm}$	47
$38\text{ cm} < w \leq 40\text{ cm}$	25
$40\text{ cm} < w \leq 42\text{ cm}$	5
$42\text{ cm} < w \leq 44\text{ cm}$	4

(a) Copy and complete this cumulative frequency table.

Waist-to-knee length w (cm)	Cumulative frequency
$32\text{ cm} < w \leq 34\text{ cm}$	
$32\text{ cm} < w \leq 36\text{ cm}$	
$32\text{ cm} < w \leq 38\text{ cm}$	
$32\text{ cm} < w \leq 40\text{ cm}$	
$32\text{ cm} < w \leq 42\text{ cm}$	
$32\text{ cm} < w \leq 44\text{ cm}$	

(b) Draw a cumulative frequency diagram of these data.

(c) Use your graph to estimate the median waist-to-knee length.

(d) Jo decides to make shorts for waist-to-knee lengths in the range 35 cm to 39 cm.
Estimate the percentage of her sample that the shorts will fit.

Answers and hints ► page 141

Standard deviation

Here are two sets of 15 measurements.

Set A: 3·2, 3·3, 3·3, 3·4, 3·6, 3·6, 3·6, 3·6, 3·7, 3·7, 3·7, 3·8, 3·8, 3·8, 3·9

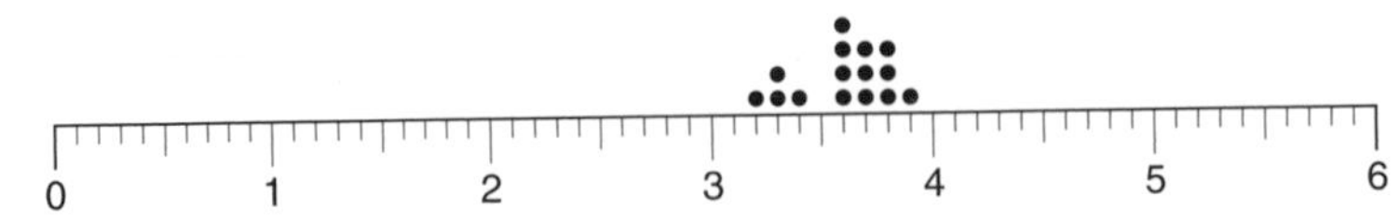

Set B: 1·5, 2·0, 3·0, 3·4, 3·5, 3·5, 3·6, 3·6, 3·6, 3·7, 3·7, 3·8, 4·8, 5·0, 5·3

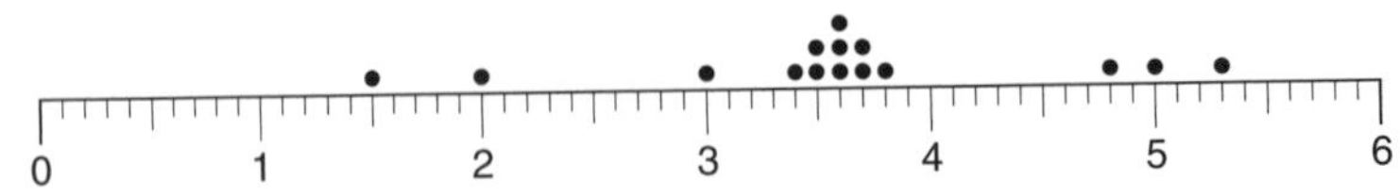

The mean value for each set is 3·6 but set B is more spread out than set A.

Range ► page 74

The **range** is one measure of spread, but it depends entirely on the two extreme values and does not tell you whether the values in between are clustered or spread out.

Inter-quartile range ► page 76

The **inter-quartile** range has a similar weakness.

The **standard deviation** measures spread using all the values. You can find it using this formula.

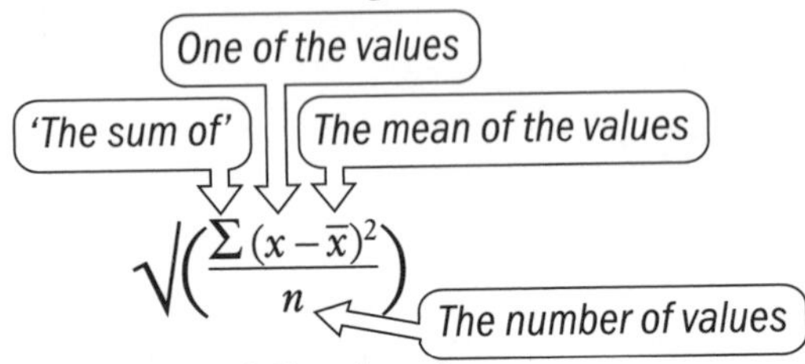

$$\sqrt{\left(\frac{\sum (x-\bar{x})^2}{n}\right)}$$

Estimating standard deviation from grouped data ► page 80

This means following these steps:

- Calculate the mean.
- Work out the difference between each value and the mean (the **deviations**).
- Square the deviations.
- Calculate the sum of the squares of the deviations.
- Divide this total by the number of values.
- Find the square root of this result.

Check that the standard deviation of set A is 0·20 and that of B is 0·96 (to 2 s.f.). This means set B is spread out about 5 times as much as set A.

Here is another formula for the standard deviation.

$$\sqrt{\left(\frac{\sum x^2}{n} - (\bar{x})^2\right)}$$

- Calculate the sum of the squares of the values.
- Divide this by the number of items.
- Subtract the square of the mean from the result.
- Find the square root of the answer.

Check that you can use this method to get the same results for sets A and B.

Your calculator may have these keys:

Key	
Σx	*the sum of all the values*
Σx^2	*the sum of the squares of the values*
$\bar{x}$	*the mean*
σ_n	*the standard deviation*

Make sure you know how to use them.

1 Calculate the mean and standard deviation for this set of data.

4·7 4·7 4·9 4·5 4·6 4·5 4·5 4·9 4·5 4·5

2 Honey is packed in jars holding 454 g.
11 samples were taken from 2 packing machines and were weighed in grams:

Machine A 447, 448, 450, 441, 452, 460, 475, 447, 454, 479, 441

Machine B 455, 450, 456, 458, 457, 452, 455, 453, 449, 453, 456

(a) Calculate the mean and standard deviation for each sample.

(b) What do these results suggest about the two machines?

3 Adam, Sven and Cindy are working in a bakery making buns.
The buns are meant to weigh 150 grams each when cooked.
The manager weighs samples of ten buns made by each of them.

	Weights of buns in grams									
Adam	152	150	154	151	157	146	149	158	154	152
Sven	152	149	153	155	152	150	150	154	152	151
Cindy	151	148	152	149	146	151	149	148	147	152

(a) Calculate the mean and standard deviation of each sample.

(b) The manager does not want complaints about buns being under-weight, but also does not want ingredients used unnecessarily.
Which worker should he be most pleased with?
Explain your reasons carefully.

4 This table shows the number of emergency calls received each day for 14 days by two ambulance stations, A and B.

	Number of emergency calls in a day													
Station A	63	58	66	56	61	48	70	61	56	71	50	53	59	66
Station B	62	63	67	60	65	65	69	68	70	67	65	62	69	64

(a) For each station, calculate the mean number of calls per day and the standard deviation.

(b) The stations have the same number of ambulances.
The local health authority plans to reduce the number of ambulances in the station that has the lower mean number of calls.
Use your answers to part (a) to comment on this proposal.

Answers and hints ► page 143

Estimates from grouped frequency data

This is a **grouped frequency table**.

Each range of values is a **class interval**.

Weight of apples in grams (x)	Frequency (f)
$90 \le x < 95$	7
$95 \le x < 100$	6
$100 \le x < 105$	9
$105 \le x < 110$	5
$110 \le x < 115$	3

To work out an **estimate of the mean** of the data, add two more columns and fill them in like this.

This is worked out from $\frac{90 + 95}{2}$.

Weights of apples in grams (x)	Frequency (f)	Mid-interval value (X)	Frequency × mid-interval value (fX)
$90 \le x < 95$	7	92·5	647·5
$95 \le x < 100$	6	97·5	585·0
$100 \le x < 105$	9	102·5	922·5
$105 \le x < 110$	5	107·5	537·5
$110 \le x < 115$	3	112·5	337·5

Total = 30   Total = 3030·0

The estimated mean is $\frac{3030{\cdot}0}{30} = 101{\cdot}0$.

To work out an **estimate of the standard deviation** you need to add this column.

Weights of apples in grams (x)	Frequency (f)	Mid-interval value (X)	Frequency × mid-interval value (fX)	Frequency × mid-interval value × mid-interval value (fX^2)
$90 \le x < 95$	7	92·5	647·5	59893·75
$95 \le x < 100$	6	97·5	585·0	57037·50
$100 \le x < 105$	9	102·5	922·5	94556·25
$105 \le x < 110$	5	107·5	537·5	57781·25
$110 \le x < 115$	3	112·5	337·5	37968·75

$\Sigma f = 30$ $\Sigma fX = 3030{\cdot}0$ $\Sigma fX^2 = 307\,237{\cdot}50$

Now use this formula:

$$\sqrt{\left(\frac{\Sigma fX^2}{\Sigma f} - \text{mean}^2\right)} \text{ which gives } \sqrt{\left(\frac{307\,237{\cdot}50}{30} - 101{\cdot}0^2\right)} = 6{\cdot}34 \text{ grams.}$$

1 The *Midland Engineering Group* employs 50 people.
The weekly wage for each employee is summarised in the table below.

Wage (£w)	Frequency	Mid-interval value	
$150 \le w < 200$	20		
$200 \le w < 250$	12		
$250 \le w < 300$	7		
$300 \le w < 350$	4		
$350 \le w < 400$	5		
$400 \le w < 450$	0		
$450 \le w < 500$	0		
$500 \le w < 550$	0		
$550 \le w < 600$	1		
$600 \le w < 650$	1		

Calculate an estimate of the mean wage of the 50 workers at the factory.
Show your working on a copy of the table.

MEG (SMP)

2 The frequency table shows the distribution of marks in a test.
Calculate an estimate of the mean and standard deviation of the marks.

Mark	Frequency
1–5	1
6–10	1
11–15	5
16–20	10
21–25	21
26–30	9
31–35	4
36–40	4
41–45	3

3 A sample of 80 eggs from a farm were weighed.
The results are shown in the table.

Mass (m grams)	Mid-interval value	Number of eggs
$40 < m \le 45$	42·5	2
$45 < m \le 50$	47·5	19
$50 < m \le 55$	52·5	24
$55 < m \le 60$	57·5	21
$60 < m \le 65$	62·5	13
$65 < m \le 70$	67·5	1

(a) Calculate an estimate of the mean mass of the 80 eggs.

(b) An egg is classified as large if it weighs more than 60 g.
  (i) What percentage of the eggs in the sample are large?
  (ii) A shopkeeper buys 2400 eggs from the farmer.
  How many would he expect to be large?

MEG (SMP)

Answers and hints ► page 144

Histograms

Make sure you understand the difference between a frequency diagram and a histogram.

This **frequency diagram** shows phone calls from an office. The **height** of this bar tells you that there were 15 calls lasting between 3 and 6 minutes (in the **class interval** 3 to 6 minutes).

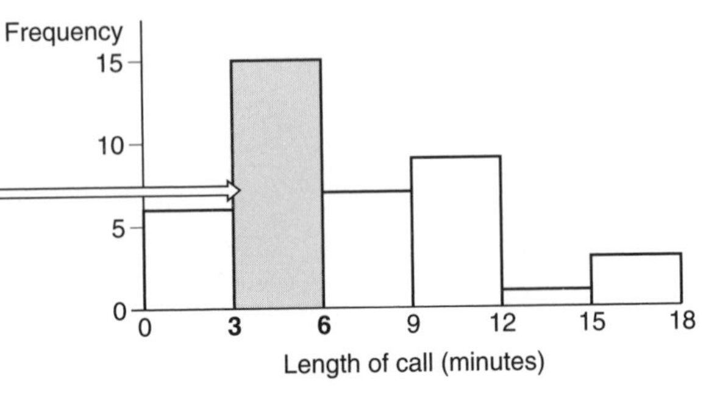

In a **histogram** the vertical axis is labelled **frequency density**.

The **area** of this bar gives you the frequency of calls in the class interval 3 to 6 minutes.

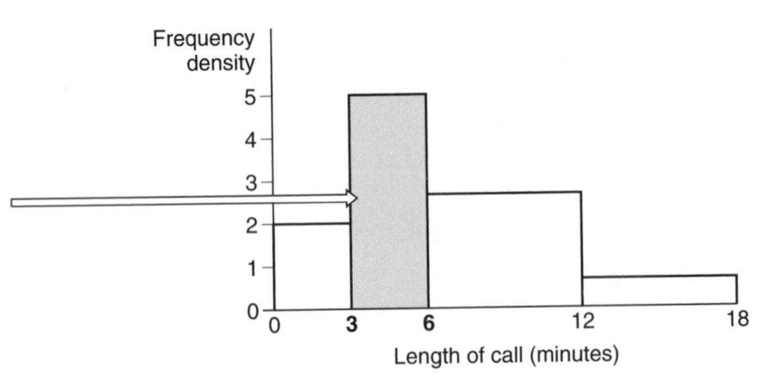

So the frequency = width of the class interval × height of the bar

$= 3 \times 5$

$= 15$

If you have to draw a histogram you work this out for each bar:

$$\text{Height of bar} = \frac{\text{frequency of items in the bar's class interval}}{\text{width of the class interval}}$$

↑ *Frequency density*

1 This histogram shows the distribution of weights of some tuna caught on a fishing trip.

(a) How many of the tuna weighed less than 20 kg?

(b) How many weighed between 20 kg and 30 kg?

(c) What was the total number of tuna caught?

(d) What percentage of them weighed 40 kg or more?

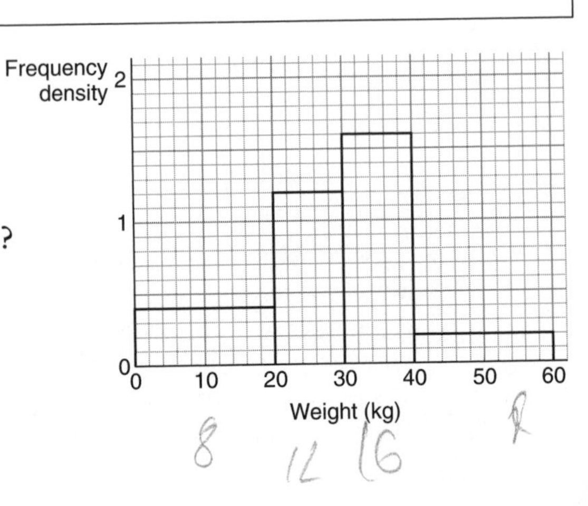

2 This histogram shows the age distribution of 78 people interviewed at a record store.

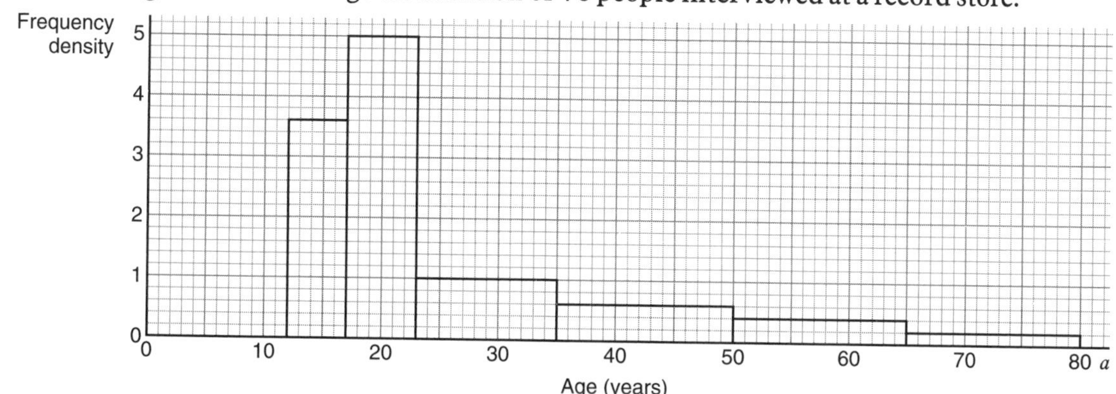

(a) Copy and complete this frequency table.

Age in years, a	Frequency
$12 \le a < 17$	
$17 \le a < 23$	
$23 \le a < 35$	
$35 \le a < 50$	
$50 \le a < 65$	
$65 \le a < 80$	

A similar survey in a garden centre gave these results.

(b) Copy this table and fill in the frequency density column.

(c) Draw the histogram for this information on worksheet H14.

(d) Comment on the differences in the distributions of ages at the two stores.

Age in years, a	Frequency	Frequency density
$12 \le a < 17$	5	
$17 \le a < 23$	3	
$23 \le a < 35$	18	
$35 \le a < 50$	42	
$50 \le a < 65$	21	
$65 \le a < 80$	16	

3 100 light bulbs were tested to see how many hours of continuous use each gave before it failed.

Use the results in the table to draw a histogram on worksheet H14.

Hours of use before failure, h	Frequency
$0 \le h < 50$	7
$50 \le h < 200$	3
$200 \le h < 800$	6
$800 \le h < 1000$	12
$1000 \le h < 1200$	58
$1200 \le h < 1400$	14

4 This table shows the number of fatal accidents for different age ranges of young people in the UK in 1992.

(a) Draw a histogram on graph paper.

(b) Which age group seems to have the highest risk of accidental death?

Age in years, a	Frequency
$0 \le a < 1$	57
$1 \le a < 5$	213
$5 \le a < 10$	176
$10 \le a < 15$	215

Answers and hints ► page 144

Approaches to sampling

If you want to find out about a large group of people (a **population**) it is easier to do a survey of just some of them (a **sample**).

To give results that are **representative** of the whole population your sample must be a reasonable size and shouldn't have any imbalance (**bias**) that would distort the result.

Random sampling

Give everyone in the population a serial number, 1, 2, 3, ... Get as many random numbers as you need for your sample size, using random number tables or a computer. Interview the people who have these random numbers as their serial numbers.

Advantage Everyone in the population has an equal chance of being in the sample.

Problems Identifying everyone in a large population to give them a serial number and then seeking out those with particular numbers takes a lot of work.
You may also get too few or too many people from a small but important group in the population (like an ethnic minority).

Systematic sampling

Interview, say, every tenth person in the population in the order you meet them.

Advantages It's a simple method and it often gives a random sample in practice.

Problem As with random sampling, you may get too few or too many people from a particular group.

Stratified sampling

Decide on all the groups that you want to represent in your sample (males and females, different age or income groups, different ethnic backgrounds, and so on).
Each group's survey results must count in proportion to its size.
So for example, if you know that 20% of the population are retired you would arrange for retired people to make up 20% of your sample.

Advantage Different groups are fairly represented.

Problems You may forget to think about an important group. You still have to decide how to get a representative sample from each group.

Quota sampling

This is an informal kind of stratified sampling. The interviewer is told, for example, 'you must interview 25 males over 40, 70 males aged 39 or under', and so on. So long as these 'quotas' are met, the interviewer uses his or her own method of selection.

Advantages It does not take a lot of work. The groups identified are fairly represented.

Problems The interviewer may unconsciously choose people of a certain type.
The time or place chosen for the interviews may cause certain groups of people to be over- or under-represented (exam questions are often about this kind of bias).

1 The local council wants to make some changes to the local road system.
The plans will cause inconvenience to some road users but
should reduce the danger to those most at risk.

The council wants to do a sample survey of people's views.
Suggest ways it could construct a stratified sample.

2 This table shows how many staff there are of different types
in an engineering company.

Proportional quantities
► **page 10**

Type	Number of employees
Management	159
Secretarial and clerical staff	428
Shop floor staff	755

A researcher wants to do a survey using a stratified sample of 150 staff.
How many staff of each type should she include?

3 A researcher for a magazine wanted to know what influenced
young people's choice of CDs.
He turned to one page in the local phone book and phoned
every tenth number on that page.

Give three reasons why this would not give a suitable sample.

4 A supermarket chain wants to test opinion on
a new range of ready meals it has just introduced.
To do this it arranges for an interviewer to question
every twentieth customer going into one of its supermarkets in
the middle of a Monday morning.

Give two reasons why this method of sampling may give misleading results.

5 There are three secondary schools in a large town.
The number of pupils in each school is given in the table.

A researcher wishes to find out what secondary school
pupils feel about the standard of education in the town.
She chooses a representative sample of total size 50.
How many pupils should be chosen from each school?

School	Pupils
Albert High School	570
St Joseph's High School	965
London Road School	1015

ULEAC

6 Katya wants to do a sample survey of residents in a particular street
to find out their views about transport.
She decides to use a systematic sample.
She knocks at numbers 10, 20, 30, and so on.

What problems might there be with this method?

Answers and hints ► page 145

Probability

If you are dealing with an experiment that has more than one stage or event, a table or tree diagram is a useful way to record the outcomes.

When the events are independent

Example

A 1 to 4 dice and a 1 to 6 dice are thrown together.
(The result on one dice does not restrict the result on the other.)
Calculate the probability of
(a) getting a total of 2
(b) getting a total of 5 or more.

Number on second dice (columns), Number on first dice (rows)

	1	2	3	4	5	6
1	(2)	3	4	5	6	7
2	3	4	5	6	7	8
3	4	5	6	7	8	9
4	5	6	7	8	9	10

The probability of getting a total of 2 is $\frac{1}{24}$ (1 possibility out of 24 equally likely ones).

The probability of getting a total of 5 or more is $\frac{18}{24}$ (18 possibilities out of 24 equally likely ones).

Example

Calculate the probability of getting just one 6 when an ordinary fair dice is thrown twice.

The dice has no memory so the probabilities for the second throw are the same as for the first.

To get these probabilities multiply together the probabilities along the branches.

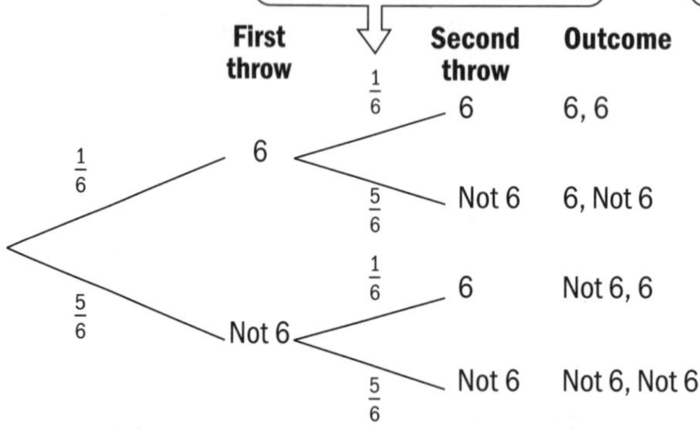

$\frac{1}{36}$, $\frac{5}{36}$, $\frac{5}{36}$, $\frac{25}{36}$

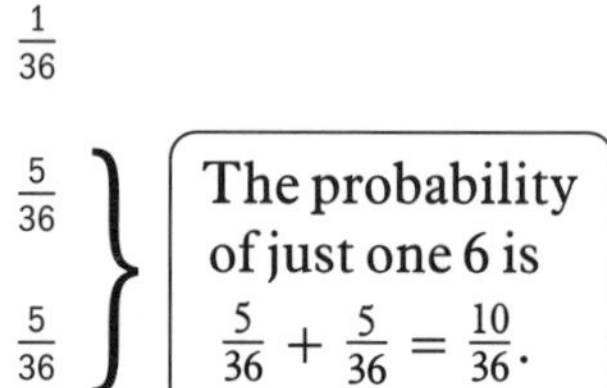

When the result for one event affects the probabilities for another event

Example

A bag contains 3 red balls and 2 blue ones.
A ball is taken out and not replaced.
A second ball is taken out.

Calculate the probability of getting just one red.

There are now 2 red and 2 blue so the probabilities change.

First ball		Second ball		Outcome	
R	$\frac{3}{5}$	R	$\frac{2}{4}$	R, R	$\frac{6}{20}$
		B	$\frac{2}{4}$	R, B	$\frac{6}{20}$
B	$\frac{2}{5}$	R	$\frac{3}{4}$	B, R	$\frac{6}{20}$
		B	$\frac{1}{4}$	B, B	$\frac{2}{20}$

The probability of just one red is $\frac{6}{20} + \frac{6}{20} = \frac{12}{20}$.

There are now 3 red and 1 blue so the probabilities change.

1 A bag contains 10 white balls, 8 blue balls and 7 yellow balls.
Pat picks one ball at random.

(a) What is the probability that she picks a blue ball?

(b) She wins a prize if she picks a blue ball or a yellow ball.
What is the probability that she wins a prize?

MEG (SMP)

2 Dinah has a biased spinner. In a trial she had these results.

Number shown	1	2	3	4	5	6
Frequency	5	16	19	20	14	26

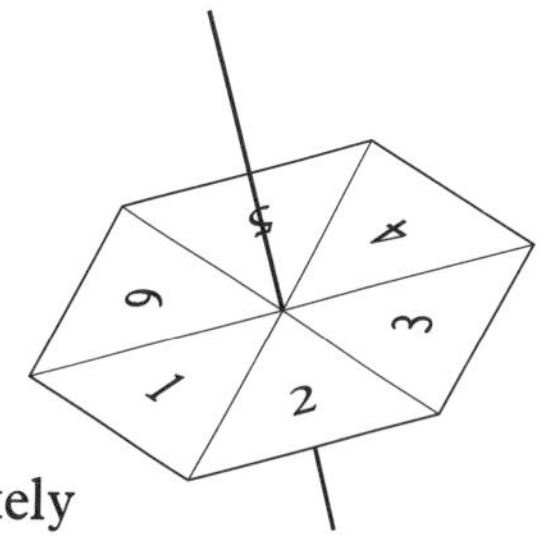

(a) (i) Use Dinah's results to estimate the probability of obtaining a six with this spinner, if she spins it once.

(ii) What should Dinah do so that she can estimate more accurately the probability of obtaining a six with this spinner?

(b) Estimate the probability of obtaining a four followed by a one with two spins.

MEG/ULEAC (SMP)

3 Ambrose has to drive through two sets of traffic lights on his way to work.
These lights work independently.
At the first set the probability he has to stop is 0·6.
At the second set this probability is 0·7.

(a) Draw a tree diagram to show this information.

(b) Calculate the probability that Ambrose is stopped by at least one of these sets of lights.

MEG (SMP)

4 Clair estimates that the probability of a drawing pin landing 'pin up' is 0·85.

(a) What is the probability of a drawing pin landing 'pin down'?

(b) Claire now drops three drawing pins.
Calculate the probability that
(i) all three land 'pin up', (ii) at least one lands 'pin down'.

MEG (SMP)

5 There are three terms in a school year.
The teaching staff have one training day in each term.
They choose each training day at random.
It is equally likely for each training day to be on any of the five days of the week from Monday to Friday.

(a) Calculate the probability that the three training days will all be on a Friday.

(b) Calculate the probability that none of the three training days will be on a Friday.

(c) Calculate the probability that the three training days will all be on the same day of the week.

(d) Calculate the probability that the three training days will be on different days of the week.

ULEAC

6 From a standard pack of playing cards a **mini-pack** is made up containing the four kings, the four queens and the four jacks.

(a) John takes two cards from this mini-pack.
Calculate the probability that they will be of the same colour.

(b) John's cards are replaced and the mini-pack is shuffled.
Jill now takes two cards from the mini-pack.
Calculate the probability that one of these will be a queen and the other will not.

(c) Jill's cards are replaced and the mini-pack is again shuffled.
Jo now takes two cards from the mini-pack.
Show that the probability that they will be of the same colour, and one will be a queen but the other will not, is $\frac{8}{33}$.

MEG/ULEAC (SMP)

7 The school canteen at St Frideswide's High School serves chips on one day a week and peas on two days a week.
These days are chosen at random in a five day week.

By drawing a tree diagram, or otherwise, calculate the probability that on Wednesday next week either chips or peas (**but not both**) will be served.

MEG (SMP)

8 The spinners in this game have been tested and found to be fair.
All three spinners are spun at the same time.

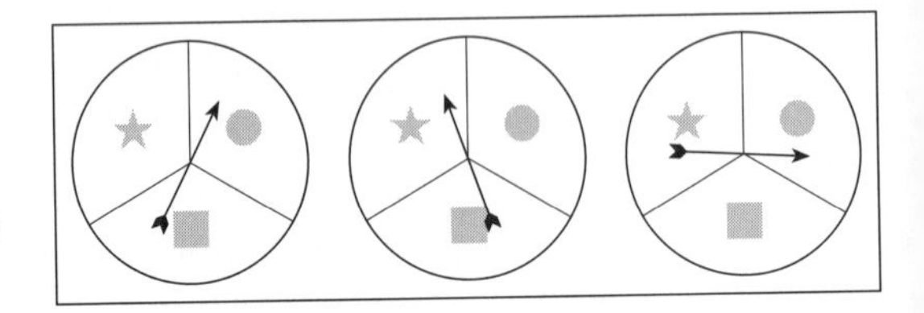

(a) What is the probability that they all point to a star when they stop?

You pay 20p to play the game.
You win £1 if all three arrows point to the same shape when the spinners stop.

(b) When you play the game, what is the probability that you win £1?

(c) The game was played 400 times at a fete.
How much money do you expect it made for the organisers, after the winnings had all been paid back?

9 There are 6 marbles in a bag, three red, two blue and one green.
Jane takes out a marble at random three times.

(a) If she replaces the marble each time, what is the probability that
(i) all three are the same colour,
(ii) all three are different colours?

(b) If she takes them out without replacement, what is the probability that
(i) all three are the same colour,
(ii) all three are different colours?

10 In the game 'Hurry home' players start with counters on the four blank corner squares.
An ordinary dice numbered 1 to 6 is thrown by each player in turn.
On each throw **all** players must move their counters to an adjoining square if the number in that square comes up on the dice.

Catherine						Bob
	●	4	5	2	●	
	5	2	1	3	5	
	4	1	6 HOME	1	2	
	3	5	1	4	3	
Damien	●	4	3	2	●	Ann

These are examples of adjoining squares.

The centre square has the number 6 and the player who reaches that square first wins the game.
If two or more players reach the centre at the same time the game is declared a draw.

(a) On the first throw of the dice what is the probability that
  (i) Ann must move,
  (ii) Ann, Bob and Catherine must all move,
  (iii) nobody has to move?

(b) What is the probability that Ann wins the game on the second throw of the dice?

(c) Bob, Catherine and Damien start a new game with their counters on the corner squares.
The first throw of the dice is a 5 and the second throw is a 4.
What is the probability that Bob wins or draws on the fourth throw of the dice?

Catherine						Bob
	●	4	5	2	●	
	5	2	1	3	5	
	4	1	6 HOME	1	2	
	3	5	1	4	3	
Damien	●	4	3	2		

NISEAC

11 The game 'Dicey' is played with two coloured dice.
The **red** dice is a standard cubical dice with faces numbered 1, 2, 3, 4, 5, 6.
The **blue** dice is a tetrahedron with faces numbered 1, 2, 3, 4.

A turn consists of these dice being thrown and the scores added according to the following rules:

1 The **red** dice is always thrown first.

2 (i) If the score on the red dice is even, then the red dice is thrown again.
  (ii) If the score on the red dice is odd, then the blue dice is used for the second throw.

Using a tree diagram, or otherwise, calculate

(a) the probability that the total score in one turn is 12,

(b) the probability that the total score in one turn is 3.

MEG (SMP)

Answers and hints ► page 146

Mixed handling data

1 The numbers of competitors finishing the 1994 London Marathon within certain times are given in the table.

Time (t hours)	Frequency	Cumulative frequency
$1.5 < t \leq 2.5$	130	
$2.5 < t \leq 3.5$	6343	
$3.5 < t \leq 4.5$	11783	
$4.5 < t \leq 5.5$	5506	
$5.5 < t \leq 6.5$	1459	
$6.5 < t \leq 7.5$	215	
$7.5 < t \leq 8.5$	31	
	25467	

(a) (i) Complete this table to show the cumulative frequencies.

(ii) On graph paper draw the cumulative frequency graph.

(b) Extend your table to calculate an estimate of the mean and standard deviation of the finish times.

(c) Use your graph to estimate the **percentage** of competitors who finished in the time intervals

(i) mean to 1 standard deviation above the mean,

(ii) mean to 1 standard deviation below the mean.

(You must indicate, on your graph, how you made the necessary readings and you must also show details of your calculations.)

MEG (SMP)

2 Zoopox is a serious illness and a reliable cure has yet to be found.

Some patients with this illness were given treatment X, some were given treatment Y, while others acted as a control group and were given no treatment.

The results are given in the table.

	Treatment X	Treatment Y	Control
Recovered	26	81	119
Died	19	54	176

(a) Find the percentage of patients receiving treatment X who recovered.

(b) Find the percentage of patients receiving treatment Y who recovered.

(c) Find the percentage of control patients who recovered.

(d) Make two comments about the treatments.

MEG (SMP)

3 (a) Fred has these four letter cards. A P Q X

How many different arrangements can he make using all four cards?

(b) Frederika has these four cards. A P P X

How can she use Fred's answer to work out the number of arrangements that she can make?

(c) Fred is given another card, S

How many different arrangements can he make using all five of his cards?

MEG (SMP)

4 A random sample of 100 packets of cereal is taken off a production line. This table shows the frequency distribution of the masses of these packets.

Mass (w grams)	Frequency
$495 \leq w < 503$	8
$503 \leq w < 507$	30
$507 \leq w < 509$	36
$509 \leq w < 513$	22
$513 \leq w \leq 521$	4

(a) On graph paper, draw a histogram to represent these data.

The mean mass of these packets is 507·4 grams.

(b) On your histogram draw and label the line $w = 507{\cdot}4$.
Hence estimate the percentage of packets that have a mass greater than 507·4 grams.

(c) Calculate, correct to 2 decimal places, an estimate for the standard deviation of these masses.

For quality control purposes, a **critical mass**, C grams, is defined by

$$C = m - 2s$$

where m is the mean, and s is the standard deviation, of the masses.

(d) Estimate the number of packets in this sample that have a mass less than C grams.

MEG/ULEAC (SMP)

5 The table shows the frequency distribution of the prices of 94 bottles of non-vintage white wines at a supermarket on 1 August 1992.

Price £p		Frequency	
$1{\cdot}75 \leq p < 2{\cdot}25$		1	
$2{\cdot}25 \leq p < 2{\cdot}75$		11	
$2{\cdot}75 \leq p < 3{\cdot}25$		24	
$3{\cdot}25 \leq p < 3{\cdot}75$		28	
$3{\cdot}75 \leq p < 4{\cdot}25$		14	
$4{\cdot}25 \leq p < 4{\cdot}75$		6	
$4{\cdot}75 \leq p < 5{\cdot}25$		3	
$5{\cdot}25 \leq p < 6{\cdot}25$		3	
$6{\cdot}25 \leq p \leq 7{\cdot}25$		4	

(a) Calculate an estimate of the mean price of these bottles of wine. (Copy the table and use the space for working if you wish.)

(b) Draw a cumulative frequency graph for the data.

(c) Use your graph to find the median price of these bottles of wine. Show your method.

(d) Which of the median and the mean would you use as the average price of these bottles of wine? Give a reason for your answer.

(e) What percentage of these bottles of wine are priced under £3?

MEG/ULEAC (SMP)

Answers and hints ► page 149

MIXED AND ORALLY-GIVEN QUESTIONS

Mixed questions 1

1 ABC is a circle, centre O, radius 10 cm.
VA and VC are tangents to this circle.
VO = 40 cm, and angle VAO = 90°.

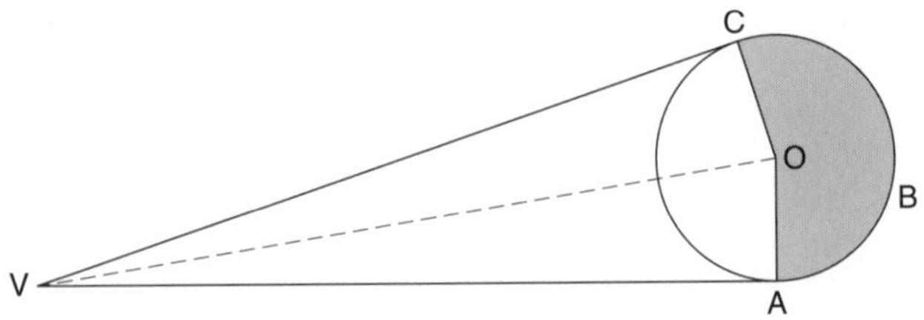

(a) Calculate (i) the length of VA,
(ii) the size of angle AOV,
(iii) the area of the sector OABC (shown shaded on the diagram).

(b) A second circle, centre P, radius r cm, is drawn to touch VA, VC and the first circle, as shown.
(i) Explain why VPO is a straight line.
(ii) Find the length of VP in terms of r.
(iii) Show that the triangles VFP and VAO are similar.
(iv) Find the value of r.

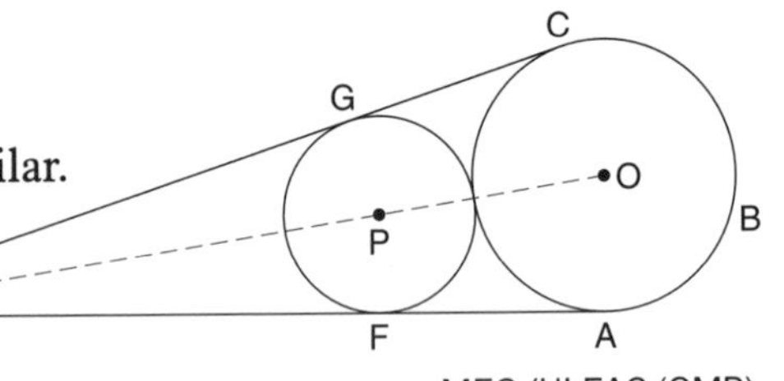

MEG/ULEAC (SMP)

2 After a storm, the rate of flow of water in a mountain stream was measured hourly.
The results are shown below.

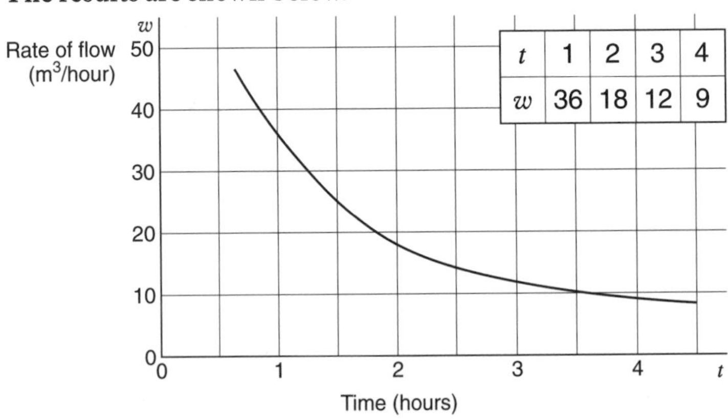

t	1	2	3	4
w	36	18	12	9

(a) Verify that, for $1 \leq t \leq 4$, w is inversely proportional to t.

(b) (i) Estimate, in cm^2, the area under the graph between $t = 1$ and $t = 4$.
(ii) What information can you use this answer to give?

(c) (i) Find the gradient of the tangent to the curve at $t = 2$.
(ii) What information is given by this gradient?

MEG (SMP)

3 (a) Factorise $x^2 - y^2$.

(b) Write 9996 as the difference of two square numbers.
Hence express 9996 as a product of prime factors.

4 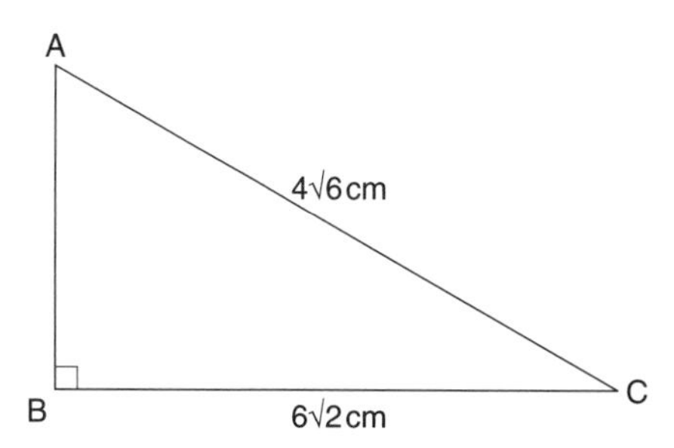

(a) Calculate the length of AB. Give your answer in its simplest surd form.

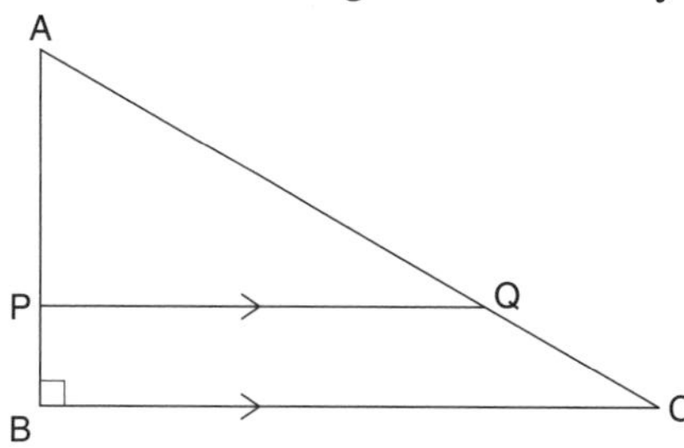

(b) In the triangle ABC given in part (a) a line PQ is drawn parallel to BC.
PQ divides the area of triangle ABC in half.
Calculate the length of AP. Give your answer in its simplest surd form. NICCEA

5 (a) State whether the following numbers are rational or irrational.
(i) $5{\cdot}252525$ (ii) $5{\cdot}\dot{2}\dot{5}$

(b) State, with reasons, whether the following lengths are rational or irrational.

(i) The circumference of a circle, radius 2 cm.

(ii) The hypotenuse length of a right-angled triangle whose other two sides are $\sqrt{5}$ cm and 2 cm.

(iii) The perpendicular height of an equilateral triangle of side 2 cm.

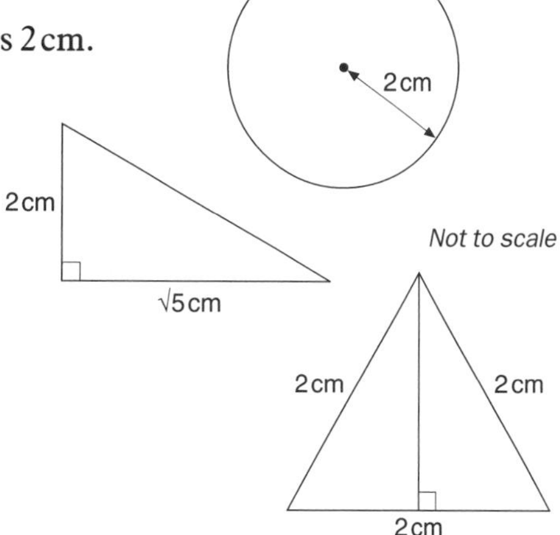

MEG

6 (a) $16^{\frac{1}{2}}$ 4^{-1} 32^0 $(^-2)^4$ 2^{-2}
Two members of the list simplify to give the same value.
Write down that value.

(b) Find the exact value of $36^{-\frac{1}{2}}$.

(c) Find the value of x if $4^x = 8$. MEG

Answers and hints ► page 151

Mixed questions 2

1 Similar right-angled triangles with sides of length a, b, c and ka, kb, kc, as shown on the right, can be placed together to form a third triangle, as shown below.

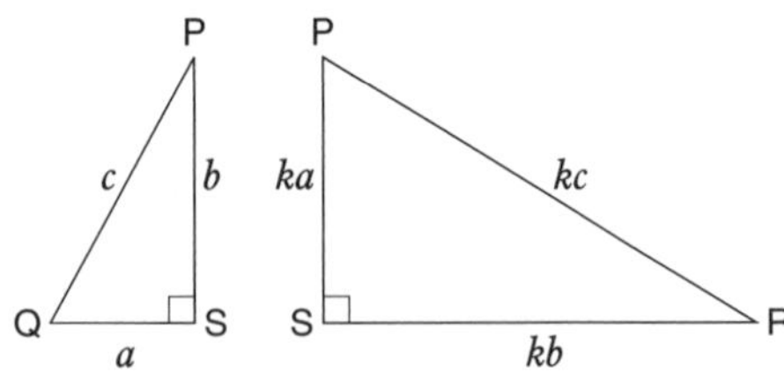

(a) Explain why angle QPR is a right-angle.

In the third triangle, RS = 3SQ.

(b) (i) Write down two equations involving a, b and k.
(ii) Hence show that $k^2 = 3$.

(c) Show that RQ = $2c$.

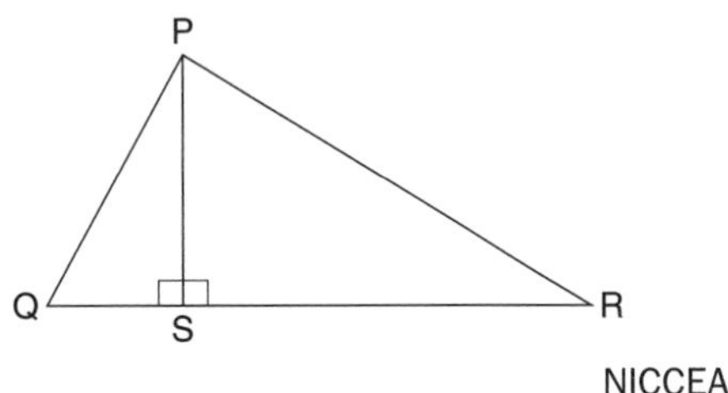

NICCEA

2 (a) Mr and Mrs Murphy enquire at two building societies about a mortgage to buy a house.

(i) The Chesterfield Building Society will allow them to borrow a total of £56 250, the equivalent of two and half times their combined annual salaries.
Representing Mrs Murphy's salary by £x and Mr Murphy's salary by £y, show that

$$x + y = 22500$$

(ii) The Countrywide Building Society will allow them to borrow a total of £58 500, the equivalent of three times Mrs Murphy's salary plus twice Mr Murphy's salary.
Write down another equation involving x and y.

(iii) Solve the two equations to find the individual salaries.

(b) When selling a house an estate agent charges a fee of $\frac{1}{2}$% of the cost of the house.
Given that the fee is £271·25, calculate the price of the house.

MEG

3 Harold makes a bow out of a piece of string of length 90 cm and a willow branch of length 60 cm. When the arrow is drawn back the branch forms a circular arc as shown. O is the centre of the circle.

Calculate the angle x.

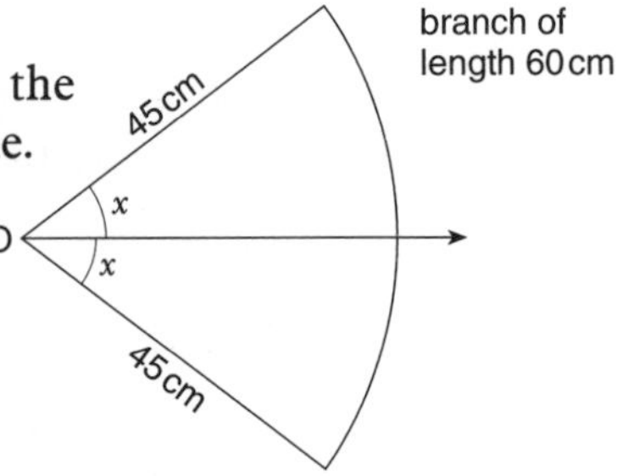

4 Sebastian runs the 100 metres race in a time given as 12·6 seconds. The track is measured correct to the nearest metre and the time is correct to the nearest 0·1 second.

What is his greatest possible average speed in metres per second?

MEG (SMP)

5 Look at the diagram alongside and identify the point which is the image of the point P after

(a) a translation $\begin{bmatrix} 0 \\ -2 \end{bmatrix}$,

(b) a reflection in the line m,

(c) an anti-clockwise rotation of 90° about O,

(d) an enlargement, centre E, scale factor 2.

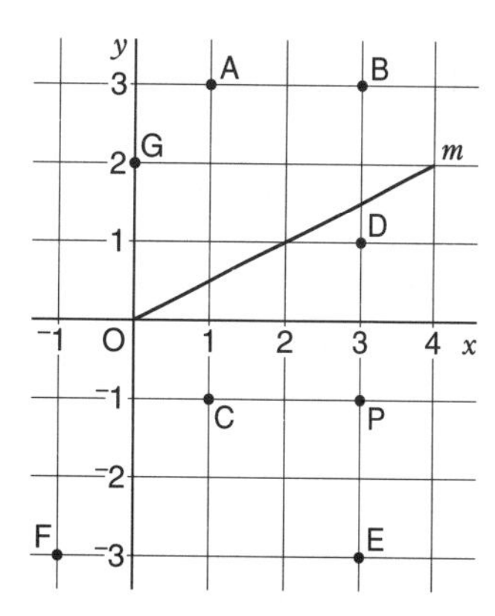

MEG (SMP)

6

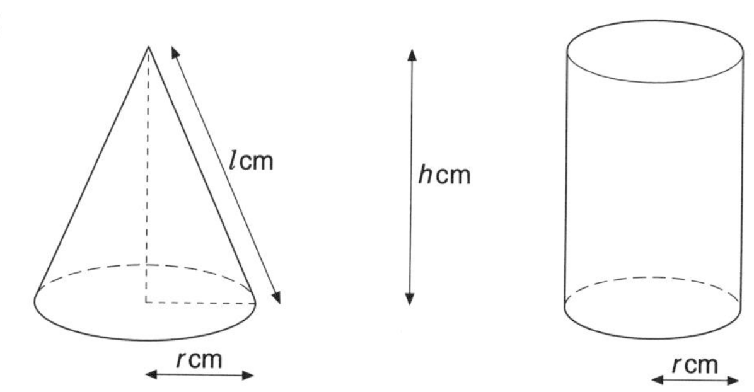

A cone with base radius r cm and height h cm has the same curved surface area as a cylinder of radius r cm and height h cm.

(a) Show that $r^2 = 3h^2$.

(b) Write down the volumes of the two solids in terms of h only.

(c) Given that the volumes differ by 16π cm^3,
  (i) show that $h = 2$,
  (ii) find r.

(d) Calculate
  (i) the slant height, l cm, of the cone,
  (ii) the curved surface area of the cone.

(e) The curved surface of the cone was made from the shaded sector of a cardboard circle, as shown.
The radius of this circle is the slant height, l cm, of the cone.
Use your answer to part (d) (ii) to express the area of the shaded sector as a fraction of the area of the complete circle.

(f) The angle of the shaded sector is x. Calculate x.

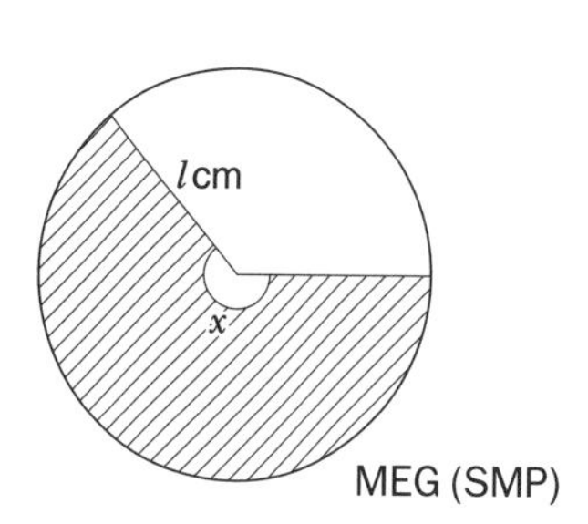

MEG (SMP)

Answers and hints ► page 152

Mixed questions 3

1 In an experiment a liquid cooled from 5 °C to $^-4$ °C in 12 minutes.

(a) By how many degrees had the temperature fallen?

(b) The temperature then fell a further 6 °C at the same rate.

(i) What was the new temperature?

(ii) How long did this further cooling take?

2 A wire in an electrical cable is a long thin cylinder.
The radius is $3{\cdot}5 \times 10^{-4}$ m and the length is $5{\cdot}8 \times 10^{3}$ m.
Work out the volume

(a) in standard form, (b) as an ordinary decimal.

MEG/ULEAC (SMP)

3 The shaded area shows a flower bed ABCD, where AB and CD are arcs of circles centre O.
The large circle has radius 3·5 m and the small one 1·5 m.

(a) Wire netting is placed all round the edge of the flower bed.
What length of wire netting is used?

(b) Find the area of the flower bed.

(c) The flower bed needs fertilising.
The manufacturer recommends 135 g of fertiliser per square metre.
What weight of fertiliser should be used for this flower bed?

MEG/ULEAC (SMP)

4 During October, a central heating boiler was used, on average, for 4·5 hours per day.
During November, the same boiler was used, on average, for 5·8 hours per day.

(a) Calculate the total time for which the boiler was used during October and November.

(b) While it is being used the boiler uses 2·44 litres of oil per hour.
Calculate the amount of oil used during the two months, correct to the nearest litre.

(c) 900 litres of heating oil cost £173. Calculate:

(i) the total cost of the oil used for the two months correct to the nearest pound,

(ii) the mean (average) daily cost of oil used over the two months correct to the nearest penny.

NICCEA

5 A rectangular carpet is placed centrally on the floor of a room 6 metres by 4 metres.
The distance from the edges of the carpet to the walls is x metres.
The carpet covers half the area of the floor.

(a) Show that $x^2 - 5x + 3 = 0$.

(b) Solve the equation in (a) to find x, correct to 3 significant figures.

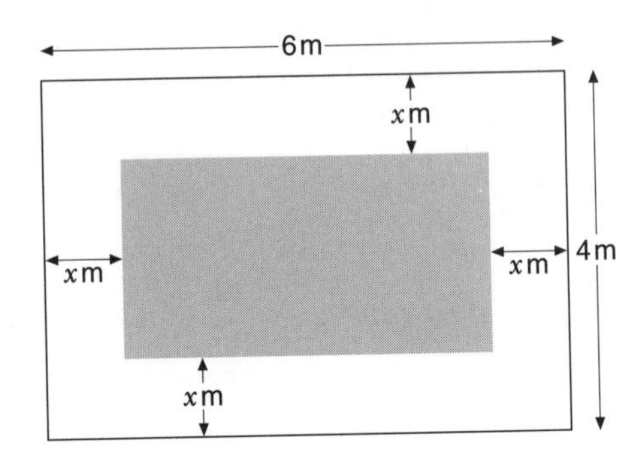

ULEAC

6 (a) Research shows that the median age, A years, of women when they marry is given by the formula

$$A = 0{\cdot}08t + 19{\cdot}8$$

where t is the number of years since 1950.

(i) Find A when $t = 20$.
(ii) Find A for the year 1993.
(iii) What was A for the year 1950?
(iv) If this trend continues, for which year will A be exactly 25?

(b) The heights of a group of boys were measured.
The results are represented in the cumulative frequency diagram below.

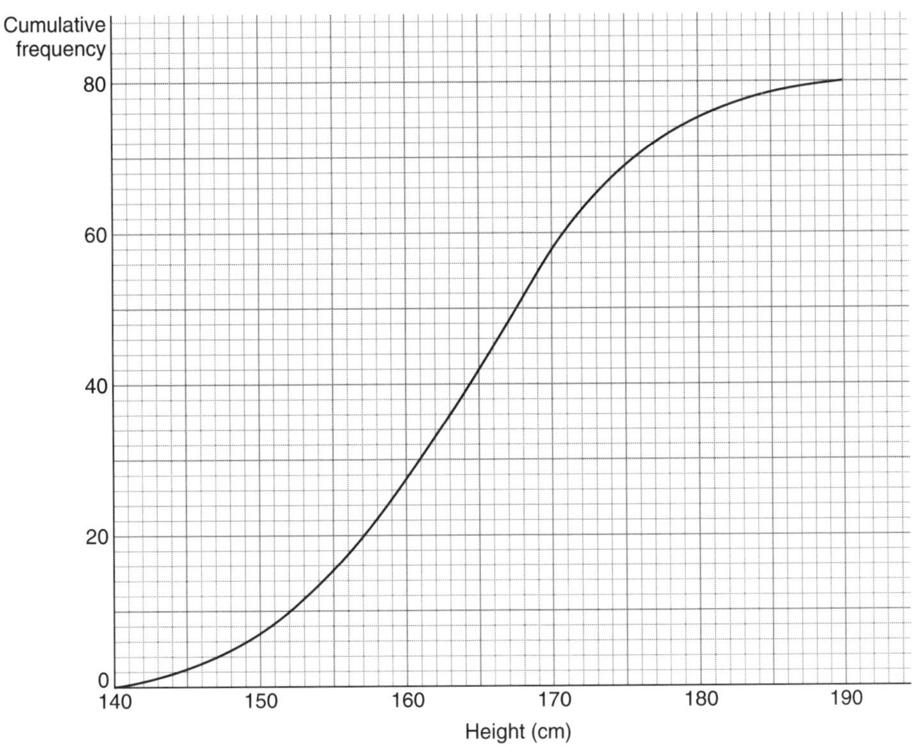

(i) What is the median height of the group?
(ii) What is the lower quartile?
(iii) What is the inter-quartile range?
(iv) Copy and complete the frequency table below for this group of boys.

Height (h cm)	$140 < h \leq 150$	$150 < h \leq 160$	$160 < h \leq 170$	$170 < h \leq 180$	$180 < h \leq 190$
Frequency					

(v) Calculate an estimate of the mean height of the group. MEG

Answers and hints ► page 153

Orally-given questions 1

Some GCSE examinations have a short test where the questions are read out to you. Practise by asking someone to read these questions to you. He or she should read each question twice and then allow the appropriate amount of time.

Do not do any working for these questions.
You will be allowed ten seconds to answer each one.

1 What is the cube root of one hundred and twenty-five?

2 Write down the next prime number after nineteen.

3 Divide twenty by nought point five.

4 What is the perimeter of a regular octagon of side two and a half centimetres?

5 Multiply six hundred by seventy.

6 Write the fraction eight twenty-fifths as a percentage.

7 What is the difference between one thousand and one, and four hundred and seven?

You may use a pencil and paper for working out these questions.
You will be allowed thirty seconds to answer each one.

8 What is the nth term of the sequence one, four, seven, ten, and so on?

9 Write the fraction eleven twentieths as a decimal.

10 What is my average speed in miles per hour if I travel eighty-five miles in two and a half hours?

11 How many CDs at fourteen pounds ninety-nine pence each can be bought for two hundred pounds?

12 Calculate the interior angle of a regular pentagon.

13 How much VAT do I pay on a television costing two hundred and fifty pounds before VAT, if the VAT rate is seventeen and a half per cent?

14 I think of a number, square it and subtract four. The answer is sixty. What number am I thinking of?

15 Write the decimal number nought point three six as a fraction in its lowest terms.

16 Peter's patio measures twenty-two feet by thirty feet. Roughly, what is the area of the patio in square metres?

17 Rob, Sue and Tariq share twenty-seven pounds in the ratio one to two to three. How much does Sue get?

18 Sketch the graph of the equation y equals one over x for positive values of x.

Answers and hints ► page 153

Orally-given questions 2

**You are not allowed to do any working for these questions.
You will be allowed ten seconds to answer each one.**

1 Write down the smallest number which is divisible by two, three and five.

2 What is the square root of one hundred and twenty-one?

3 If the temperature falls from twelve degrees Celsius to minus seven degrees Celsius, what is the change in degrees Celsius?

4 Add four tenths to five point eight.

5 The diameter of a wheel is eighty centimetres.
Roughly what is its circumference?

6 Write down the reciprocal of one fifth.

7 How many lines of symmetry does a regular hexagon have?

**You may use a pencil and paper for working out these questions.
You will be allowed thirty seconds to answer each one.**

8 One angle of an isosceles triangle is ninety-four degrees.
What are the other two angles?

9 Under a translation, the image of the point (3, 4) is ($^-1$, 7).
Write down the column vector of the translation.

10 Four friends share equally the cost of hiring a car.
How much do they each pay if the total cost is seventy-seven pounds and forty-eight pence?

11 What is the probability of throwing a multiple of two with an unbiased nought to five dice?

12 What is the nth term of the sequence one, four, nine, sixteen, and so on?

13 If the bearing of B from A is one hundred and thirty-five degrees, what is the bearing of A from B?

14 Sketch the graph of y equals x cubed for positive and negative values of x.

15 Paul buys ten concert tickets costing nine pounds fifty pence each and gets a discount of ten per cent. How much does he pay altogether?

16 An albatross was tracked by satellite and found to travel fifteen thousand kilometres in thirty days.
On average, how far did it travel in a single day?

17 Write down this inequality: five plus the product of two and x is greater than the value of x cubed.

18 Estimate the answer to five hundred and eighty divided by the square of twenty-nine.

Answers and hints ► page 154

ANSWERS AND HINTS

Accuracy of answers

The accuracy of your answers is important. You will often have to decide for yourself on a suitable degree of accuracy; if you do not, you could lose a mark. In general, you should never give an answer to more accuracy than the data in the question and sometimes one figure less might be appropriate.

In multi-stage calculations you may need to give a rounded answer to one part of a question. If you use this answer in another part, be sure to use the unrounded answer in your calculation.

NUMBER

Properties of numbers (page 2)

1 (a) 1, 2, 4, 7, 14 and 28
(b) $2 \times 2 \times 7$ or $2^2 \times 7$
(c) 41, 43, 47, 53, 59, 61, 67
(d) 12, 24, 36, ...

2 (a) 0·25 (b) 27000 (c) 12
(d) 10 (e) 0·008 (f) 10

3 (a) $\frac{1}{20}$ or 0·05 (b) 5 (c) $\frac{1}{400}$ or 0·0025
(d) $\frac{3}{2}$ or 1·5 (e) $\frac{1000}{5} = 200$
(f) $\frac{1}{1000}$ or 0·001 or 10^{-3}

4 (a) 5, 13, 37, 47 (b) 80
(c) 1, 25, 64 (d) 1 and 64

5 13

More help or practice
Reciprocals and division ► Book Y1 pages 6 to 8

Negative numbers and fractions (page 3)

Remember that the [+/–] *key on a calculator changes the sign, so to key in* $^-3$*, you press* [3] [+/–].

1 (a) $^-3$ (b) 11 (c) $^-36$ (d) 3
(e) $^-4{\cdot}5$ (f) 5 (g) $^-10$ (h) $^-0{\cdot}8$

2 (a) $(9 - 15)°\text{C} = {}^-6\,°\text{C}$
(b) $({}^-6 - 4)°\text{C} = {}^-10\,°\text{C}$

3 $^-$£89·50 + £250 + £21·75 – £85·34 = £96·91

4 (a) $\frac{2}{3} + \frac{3}{5} = \frac{10}{15} + \frac{9}{15} = \frac{19}{15} = 1\frac{4}{15}$

(b) $1\frac{5}{8} - \frac{2}{3} = \frac{13}{8} - \frac{2}{3} = \frac{39}{24} - \frac{16}{24} = \frac{23}{24}$

(c) $\frac{2}{5} \times \frac{3}{4} = \frac{2 \times 3}{5 \times 4} = \frac{6}{20} = \frac{3}{10}$

$\frac{6}{20}$ is simplified by dividing both numerator (top number) and denominator (bottom number) by 2.

(d) $\frac{9}{10} + \frac{3}{5} - \frac{3}{8} = \frac{9+6}{10} - \frac{3}{8} = 1\frac{1}{2} - \frac{3}{8} = 1\frac{1}{8}$

or

$\frac{9}{10} + \frac{3}{5} - \frac{3}{8} = \frac{36 + 24 - 15}{40} = \frac{45}{40} = 1\frac{5}{40} = 1\frac{1}{8}$

40 is the lowest common denominator of 10, 5 and 8; that is, the smallest number of which they are all factors.

(e) $\frac{3}{4}$ of $1\frac{1}{2} = \frac{3}{4} \times \frac{3}{2} = \frac{9}{8} = 1\frac{1}{8}$

(f) $\frac{7}{10} \div 5 = \frac{7}{10} \times \frac{1}{5} = \frac{7}{50}$

(g) $2\frac{1}{2} \div \frac{2}{3} = \frac{5}{2} \times \frac{3}{2} = \frac{15}{4} = 3\frac{3}{4}$

(h) $\frac{1}{2}(\frac{1}{4} \times \frac{2}{5}) = \frac{1}{2} \times \frac{1}{4} \times \frac{2}{5} = \frac{1 \times 1 \times 2}{2 \times 4 \times 5} = \frac{2}{40} = \frac{1}{20}$

5 $1 - \frac{3}{5} - \frac{1}{4} = \frac{10}{10} - \frac{3}{5} - \frac{1}{4} = \frac{20 - 12 - 5}{20} = \frac{3}{20}$

More help or practice
Add and subtract negative numbers ► Book Y1 pages 57 to 59
Multiply and divide directed numbers ► Book Y1 pages 60 to 62
Add and subtract fractions ► Book R+ page 52, Book Y4 page 82
Multiply fractions ► Book R+ page 60

Getting rough answers; using a calculator (page 4)

1 (a) 4·09 (b) 6·0 (c) 0·010
In (c) the '9' is rounded up to '10'; the right-hand zero is the third decimal place.

2 (a) 3800 (b) 0·078 (c) 0·070
(d) 3·0 (e) 330 (f) 1
The answer to (f) would be 1·0 if 2 significant figures were asked for.

3 *Here is one way of doing the estimation; there are others. The important thing is to show your working. (We have rounded 0·92 to 1 instead of 0·9 to keep the arithmetic simple.)*

$$\frac{680^2}{53 \times 0{\cdot}92} \approx \frac{700^2}{50 \times 1} = \frac{49000}{50} \approx 1000$$

4 $$\frac{15{\cdot}52 \times 0{\cdot}913}{102{\cdot}8 \times 0{\cdot}0314} \approx \frac{16 \times 1}{100 \times 0{\cdot}03} = \frac{16}{3} \approx 5$$

5 *One method of checking by rounding is given after each answer.*

(a) $0{\cdot}50715\ldots = 0{\cdot}51$ (to 2 s.f.)
$[60 \div (50 + 70) = 60 \div 120 = 0{\cdot}5]$

(b) $28038{\cdot}1\ldots = 28000$ (to 2 s.f.)
$[(9 \times 8) \div 0{\cdot}05^2 = 72 \div 0{\cdot}0025 \approx 28000]$

(c) $1{\cdot}1458\ldots = 1{\cdot}1$ (to 2 s.f.)
$[(0{\cdot}4 \times 20) \div 6 = 8 \div 6 \approx 1{\cdot}3]$

(d) $10{\cdot}4803\ldots = 10$ (to 2 s.f.)
$[8^2 \div 6 = 64 \div 6 \approx 11]$

6 (a) (i) 14·03566885
(ii) 14·0 or 14

(b) $$\frac{29{\cdot}5^2 - 19{\cdot}7^2}{2 \times 9{\cdot}8} \approx \frac{30^2 - 20^2}{20} = 25$$

7 Colin pays $28{\cdot}4 \times 52{\cdot}5$p = £14·91.
At the second garage this buys
$14{\cdot}91 \div 0{\cdot}462 = 32{\cdot}2727\ldots$ litres; that is, an extra 3·9 litres (to 2 s.f.).

8 (a) Total payment
= £28·00 + (£8·40 × 9) = £103·60

(b) £5·20 × 21 = £109·20
Sandra paid £(109·20 – 103·60) = £5·60 more than Leza.

9 (a) Tim expected to receive
$(800 × 1·578) = $1262·40.

(b) Tim paid £(1300 ÷ 1·578) + £5 = £828·83 (to the nearest penny).

10 1146 pages (or 573 sheets) are 59·5 mm thick.
600 sheets would be approximately 60 mm thick, so 1 page is approximately 0·1 mm thick.

> **More help or practice**
> Rounding ► Book Y1 pages 1, 38 to 39
> Significant figures ► Book Y1 pages 40 to 41

Fractions, decimals and percentages (page 6)

1 (a) 0·84 (b) 0·9 (c) 0·02
(d) $\frac{3}{16} = 3 \div 16 = 0{\cdot}1875$ (e) 0·125

2 (a) 70% (b) 62·5% (c) 72·5%
(d) 65% (e) 240%

3 (a) $\frac{3}{4}$ (b) $\frac{3}{25}$ (c) $\frac{7}{40}$
(d) $\frac{1}{20}$ (e) $\frac{13}{200}$

4 (a) 13 kg (b) £0·75 or 75p
(c) £0·54 (d) 8·4 m

5 (a) 0·647 (to 3 d.p.)

(b) $\frac{3}{5} = 0{\cdot}6$, $\frac{11}{17} = 0{\cdot}647$ (to 3 d.p.), $65\% = 0{\cdot}65$;
so the order is $\frac{3}{5}$, 0·63, $\frac{11}{17}$, 65%.

6 $\frac{4}{11} = \frac{32}{88}$ and $\frac{3}{8} = \frac{33}{88}$, so $\frac{3}{8}$ is the larger.
You could also have converted each fraction to a decimal.

7 117·5% of £79·25 = £79·25 × 1·175
= £93·12 to the nearest penny
You could also have worked out 17·5% of £79·25 and added it to £79·25.

8 (a) Deposit = 25% of £160 = £40

(b) Total monthly payments
= £14·40 × 12 = £172·80
Total paid by hire purchase = £212·80
Extra payment = £52·80

9 (a)

	Jeep	Camel
Under 30	50·7%	**49·3%**
Over 30	**65%**	**35%**

(b) The shaded square represents the percentage of camel riders who are under 30.

10 Mary paid tax on £32600 – £3765 = £28835
Her total income tax
= £(3900 × 20% + 21600 × 24% + 3335 × 40%)
= £(780 + 5184 + 1334)
= £7298

11 *Start by completing the table.*

Membership of a National Charity

	Men	Women	Total
Live in Scotland	415678	267350	683028
Live elsewhere	73897	105951	179848
Total	489575	373301	862876

(a) $\frac{415678}{489575} \times 100\% = 84{\cdot}9\%$ (to 3 s.f.)

(b) $\frac{373301}{862876} \times 100\% = 43{\cdot}3\%$ (to 3 s.f.)

More help or practice

Percentage and decimal equivalents ► Book Y1 page 114
Expressing one amount as a percentage of another ► Book Y1 page 115
Percentages greater than 100 ► Book Y1 pages 116 to 117
Calculating percentage increases and decreases ► Book Y1 pages 118 to 121
Using percentages for comparison ► Book Y3 pages 36 to 37
Two-way tables ► Book Y3 pages 38 to 41

Percentages (page 8)

1 The increase is 14p or £0·14.
So the percentage increase
$= \frac{0{\cdot}14}{2{\cdot}25}$
= 0·06 = 6% to the nearest per cent

2 The multiplying factors are 1·13 and 0·955.
So cost = £310 × 1·13 × 0·955 = £334·54

3 1·08 × 1·15 = 1·242
So Mike's overall percentage increase is 24% (to 2 s.f.).

4 (a) Amount kept = £$\frac{4}{5} \times \frac{1}{4} \times 2\frac{1}{4}$ million
= £$\frac{4}{5} \times \frac{1}{4} \times \frac{9}{4}$ million
= £$\frac{9}{20}$ million
= £0·45 million or £450000

(b) The multiplying factors are 1·2 and 0·8.
1·2 × 0·8 = 0·96
So the percentage change is a decrease of 4%.
Note that the factor of 0·8 comes from the price reduction of 20%.

5 (a) Price reduction = £2·60
True 'percentage off' = $\frac{2{\cdot}60}{8{\cdot}99} \times 100$
= 28·9 (to 1 d.p.)
You could also have used the alternative method on page 8 to calculate this.

(b) Original price × 0·80 = £6·39
Original price = £6·39 ÷ 0·80
= £7·99
(to the nearest penny)

6 Lisa has a 35% discount on the full premium.
So 0·65 × full premium = £268·80
full premium = £268·80 ÷ 0·65
= £413·54
(to the nearest penny)

7 (a) The number of rabbits in 1994 $= 308 \times 0{\cdot}88$
$= 271$

(b) The number of rabbits in 1992 $= 308 \div 0{\cdot}88$
$= 350$

(c) *You need to use successive multiplication by 0·88 to find the number of years for the population to fall below 180 rabbits.*
$271 \times 0{\cdot}88 \times 0{\cdot}88 \times 0{\cdot}88 \times 0{\cdot}88 = 162{\cdot}517$
The required year is 1998.

In questions like this always be careful to think about whether you are dealing with increases or decreases and do not round until the end of the calculation.

8 (a) Commission $= £300 \times 0{\cdot}015$
$= £4{\cdot}50$
Bernard receives $295{\cdot}5 \times 8{\cdot}4$ francs
$= 2482{\cdot}2$ francs.

(b) The amount needed after paying 1·5% commission is $£\frac{1500000}{2320}$.

So actual cost $= £\frac{1500000}{2320 \times 0{\cdot}985}$
$= £656{\cdot}40$ (to nearest penny).

More help or practice
Percentages of percentages ► Book Y3 pages 92 to 95
Compound interest ► Book Y3 pages 95 to 97

Ratio and proportion (page 10)

1 (a) 4 : 3 (b) 1 : 3 (c) 35 : 100 = 7 : 20
(d) 60 : 12 = 5 : 1

Did you remember to change £1 to pence in (c) and 1 hour to minutes in (d) so that you had the same units?
Ratios in simplest form must not include fractions.

2 The volume of oil is given by
$(800 \div 25)$ ml $= 32$ ml.

3 (a) Anne receives $\frac{12}{20} \times 200\text{p} = 120\text{p}$ or £1·20.

Bobby receives $\frac{8}{20} \times 200\text{p} = 80\text{p}$ or £0·80.

Remember to check that the two amounts sum to the total, in this case £2.

(b) Two years ago Anne was 10 and Bobby was 6 years old.
The ratio of their ages was 10 : 6 = 5 : 3.
Bobby received 30p so Anne received 50p.

4 9p

5 (a) £75

(b) The amount charged is not proportional to the time taken because of the call-out fee; for example, the cost of a 2-hour visit is not double the cost of a 1-hour visit.

6 (a) The proportion of copper $= \frac{185}{185 + 14 + 1} = \frac{185}{200}$

The amount of copper $= \frac{185}{200} \times 120$ kg
$= 111$ kg

(b) The proportion of tin $= \frac{14}{200} = \frac{7}{100}$

If x kg is the total weight of alloy then

$$\frac{7}{100}x = 4{\cdot}9$$
$$x = \frac{100}{7} \times 4{\cdot}9$$
$$= 70$$

So the total weight of alloy is 70 kg.

7 Using the multiplier method, the 'perf' is given by $27 \times \frac{2}{3{\cdot}6} = 15$.

8 Using the unitary method, the pressure is given by

$$\frac{1{\cdot}8 \times 10^5}{288} \times 303\,\text{N/m}^2 = 1{\cdot}89375 \times 10^5\,\text{N/m}^2$$
$$= 1{\cdot}89 \times 10^5\,\text{N/m}^2 \text{ (to 3 s.f.)}.$$

'K' stands for 'kelvin', a temperature scale used in physics.

More help or practice
The unitary method ► Book Y1 pages 43 to 48
The multiplier method ► Book Y1 pages 71 to 73, Book Y2 pages 103 to 105
Ratios less than 1 ► Book Y1 page 74
'No change' rule ► Book Y1 page 75
Non-proportionality ► Book Y1 page 106

Types of proportionality (page 12)

1 (a)

x	6	24	48	120
y	6·25	100	400	2500

(b)

x	6	24	48	120
y	400	100	50	20

2 (a) *Substitute known values in* $y = \frac{k}{x^2}$ *to find the value of* k.

$10 = \frac{k}{4}$

$k = 40$

So $y = \frac{40}{x^2}$ or $yx^2 = 40$

(b) (i) $y = 2{\cdot}5$ (ii) $x = 1$ or $^{-}1$

3 (a) *Substitute a pair of values in* $y = kx^2$ *to find the value of* k.

$45 = 30^2k = 900k$

$k = \frac{45}{900} = \frac{1}{20}$

So $y = \frac{x^2}{20}$ or $20y = x^2$

(b) When $x = 75$, $20y = 75^2 = 5625$

$y = 281{\cdot}25$

The braking distance is 280 feet (to 2.s.f.).

(c) When $y = 400$, $20 \times 400 = x^2$

$x^2 = 8000$

$x = 89{\cdot}4$

The speed is 89 m.p.h. (to 2 s.f.).

4 The density is reduced by a factor of 1·5 (or equivalent answer).

5 $1{\cdot}6 = k \times 2{\cdot}5^2$

$k = \frac{1{\cdot}6}{2{\cdot}5^2}$

Energy stored in second battery

$= \frac{1{\cdot}6}{2{\cdot}5^2} \times 1{\cdot}5^2 = 0{\cdot}576$ units of energy

It is usually best to leave all the calculation to the end, especially if rounding is involved.

6 (a)

t	3	30	15	5	50	100
d	180	18000	4500	500	50000	200000
F	1000	10	40	360	3·6	0·9

(b) *Start by calculating the constant of proportionality,* k.

(i) $d = kt^2$

$k = \frac{180}{9} = 20$

$d = 20t^2$

(ii) $F = \frac{k}{d}$

$k = Fd$

$= 180000$

$Fd = 180\,000$ or $F = \frac{180\,000}{d}$

7 (a) The constant of proportionality is $20 \times 16{\cdot}5 = 330$.

The frequency for a 1 m wavelength is $330 \div 1 = 330$ Hertz.

(b) Wavelength $= \frac{330}{15000} = 0{\cdot}022$ metres

8 *Use the pair of values you know to find the constant of proportionality,* k, *and then use this to find* s *and* t *by substitution.*

$k = \frac{108}{6^3} = 0{\cdot}5$

$Q = kP^3 = 0{\cdot}5P^3$

So $s = 0{\cdot}5 \times 0{\cdot}8^3 = 0{\cdot}256$

Similarly, $13{\cdot}5 = 0{\cdot}5t^3$, so

$t^3 = 13{\cdot}5 \div 0{\cdot}5 = 27$ and

$t = \sqrt[3]{27} = 3$

More help or practice

Direct and inverse proportionality
(using the multiplier rule and constant ratio rule)
► Book Y3 pages 68 to 74, Book Y4 pages 62 to 63

Other types of proportionality ($p \propto q^2, p \propto q^3, p \propto \frac{1}{q^2}$)
► Book Y4 pages 64 to 67

Indices (page 14)

1 (a) 6600 (b) 0·00007 (c) 0·0483
(d) 4 (e) 0·1 or $\frac{1}{10}$

2 (a) $5{\cdot}23 \times 10^5$ (b) $4{\cdot}6 \times 10^{-5}$
(c) $1{\cdot}007 \times 10^{-2}$ (d) $7{\cdot}811 \times 10^7$
(e) 1×10^{-4}

3 (a) 5^7 (b) 3^{10} (c) $7^{\frac{4}{3}}$ (d) 8^2 or 2^6
(e) 3^{-2} (f) 11^{-4} (g) 2^{-4} (h) $4^{\frac{3}{2}}$ or 2^3 or 8

4 (a) 2^3 (b) 2^{-2} (c) 2^9 (d) 2^{-6} (e) $2^{-\frac{1}{2}}$

5 (a) 7^2 (b) 3^3 (c) 2^{-3}
(d) $3^{\frac{3}{2}}$ (e) $2^{\frac{2}{3}}$ (f) $\frac{1}{25} = 5^{-2}$

6 (a) 9765625 (b) 2097152
(c) 3·68403… (d) 2·44140…
(e) 0·04110…

7 (a) $(6{\cdot}31 \div 4{\cdot}52) \times 10^{13} = 1{\cdot}39601\ldots \times 10^{13}$
$= 1{\cdot}40 \times 10^{13}$ (to 3 s.f.)
(b) $(2{\cdot}31 \div \sqrt{5{\cdot}56}) \times 10^9 = 0{\cdot}97965\ldots \times 10^9$
$= 9{\cdot}80 \times 10^8$ (to 3 s.f.)

8 $1{\cdot}5 \times 10^{101}$

9 Total daily amount $= 1{\cdot}8 \times 1{\cdot}4 \times 10^7$ kg
Annual amount $= 1{\cdot}8 \times 1{\cdot}4 \times 10^7 \times 365$ kg
$= 919{\cdot}8 \times 10^7 \times 10^{-3}$ tonnes
$= 9{\cdot}2 \times 10^6$ tonnes (to 2 s.f.)
Show all your working so that if you make an error in calculation you can still gain method marks.

10 (a) $9{\cdot}5 \times 10^{12}$
(b) $9{\cdot}5 \times 4{\cdot}3 \times 10^{12} = 40{\cdot}85 \times 10^{12}$
$= 4{\cdot}1 \times 10^{13}$ (to 2 s.f.)
(c) $4{\cdot}085 \times 10^{13} \div (1{\cdot}5 \times 10^8) = 2{\cdot}72333\ldots \times 10^5$
$= 2{\cdot}7 \times 10^5$ (to 2 s.f.)

11 (a) $3{\cdot}45 \times 10^{-4}$
(b) $\frac{1}{A} = 0{\cdot}2 \times 10^8 = 2 \times 10^7$
(c) $(2{\cdot}52 \times 10^7) \div (9{\cdot}67 \times 10^5) = 0{\cdot}26059\ldots \times 10^2$
$= 26{\cdot}1$ (to 3 s.f.)

More help or practice
Multiply and divide with positive indices
► Book YR+ pages 29 to 30
Negative indices and standard form ► Book Y4 pages 96 to 97
Estimate answers to problems involving standard form
► Book Y5 pages 114 to 115
Manipulate fractional and negative indices
► Book YX2 pages 27 to 31

Rational and irrational numbers (page 16)

1 (a) $\frac{4}{11}$ (b) $\frac{11}{9}$ or $1\frac{2}{9}$ (c) $\frac{4{\cdot}7}{99} = \frac{47}{990}$
(d) $\frac{55{\cdot}6}{99} = \frac{556}{990} = \frac{278}{495}$ (e) $\frac{631{\cdot}1}{999} = \frac{6311}{9990}$

2 (a) $\sqrt{300} = \sqrt{3} \times \sqrt{100} \approx 1{\cdot}73 \times 10 = 17{\cdot}3$
(b) $\sqrt{27} = \sqrt{3} \times \sqrt{9} \approx 1{\cdot}73 \times 3 = 5{\cdot}19$

3 (a) $\sqrt{8} \times \sqrt{2} = \sqrt{4} \times \sqrt{2} \times \sqrt{2} = 4$ or
$\sqrt{8} \times \sqrt{2} = \sqrt{(8 \times 2)} = \sqrt{16} = 4$
(b) $\frac{6\sqrt{2}}{\sqrt{6}} = \frac{6\sqrt{2} \times \sqrt{6}}{\sqrt{6} \times \sqrt{6}} = \frac{6\sqrt{12}}{\sqrt{6}} = 2\sqrt{3}$
(c) $(\sqrt{3} + \sqrt{2})^2 = 3 + \sqrt{6} + \sqrt{6} + 2$
$= 5 + 2\sqrt{6}$
(d) $\sqrt{48} + \sqrt{3} = (\sqrt{16} \times \sqrt{3}) + \sqrt{3}$
$= 4\sqrt{3} + \sqrt{3}$
$= 5\sqrt{3}$
(e) $\frac{\sqrt{7000}}{\sqrt{700}} = \frac{\sqrt{100} \times \sqrt{70}}{\sqrt{100} \times \sqrt{7}} = \frac{\sqrt{7} \times \sqrt{10}}{\sqrt{7}} = \sqrt{10}$
(f) $\sqrt{12} - \sqrt{3} = \sqrt{4} \times \sqrt{3} - \sqrt{3}$
$= 2\sqrt{3} - \sqrt{3} = \sqrt{3}$
So $(\sqrt{12} - \sqrt{3})^2 = (\sqrt{3})^2 = 3$
(g) $(\sqrt{5} + \sqrt{2})(\sqrt{2} - \sqrt{5}) = 2\sqrt{5} - 5 + 4 - 2\sqrt{5} = -1$
(h) $\frac{4}{\sqrt{2}} + \frac{3}{\sqrt{2}} = \frac{7}{\sqrt{2}} = \frac{7}{2}\sqrt{2}$

4 (a) 2 is rational and $\sqrt{3}$ is irrational. The product of a rational and an irrational number is always irrational, so $\sqrt{12}$ is irrational.
(b) $\sqrt{36} = 6$ which is rational. Since the product of a rational and an irrational number is always irrational, $\sqrt{7{\cdot}2}$ cannot be rational and so must be irrational.

5 (a) $3 + \sqrt{5}$; irrational

(b) $3 + \sqrt{5} + 3 - \sqrt{5} = 6$; rational

(c) $3 + \sqrt{5} - (3 - \sqrt{5}) = 2\sqrt{5}$; irrational

(d) $(3 + \sqrt{5})(3 - \sqrt{5}) = 9 - 5 = 4$; rational

6 (a) $0\cdot\dot{1}$ can be written as the rational fraction $\frac{1}{9}$.

Any recurring decimal would be acceptable.

(b) $\sqrt{(4\frac{1}{4})} = \sqrt{\frac{17}{4}} = \frac{\sqrt{17}}{2}$ and $\frac{1}{3} + \sqrt{3}$ are irrational.

$\sqrt{(6\frac{1}{4})} = \sqrt{(\frac{25}{4})} = \frac{5}{2}$ and

$(\frac{1}{3}\sqrt{3})^2 = \frac{1}{9} \times 3 = \frac{1}{3}$ are rational.

7 (a) $0\cdot\dot{3} = \frac{3}{9} = \frac{1}{3}$

(b) $4^{\frac{1}{2}} = \sqrt{4} = 2$

(c) $8^{-\frac{1}{2}} = \frac{1}{\sqrt{8}} = \frac{1}{2\sqrt{2}} = \frac{1}{4}\sqrt{2}$

(d) $3\cdot142 = \frac{3142}{1000}$

(e) $(\sqrt{2})^{-4} = \frac{1}{(\sqrt{2})^4} = \frac{1}{2 \times 2} = \frac{1}{4}$

(f) $(1 + \sqrt{5})(1 - \sqrt{5}) = 1 - 5 = {}^{-}4$

$8^{-\frac{1}{2}}$ is the only irrational number.

8 (a) (i) $p^2 - q^2 = 9 - 16 = {}^{-}7$

(ii) $p^{\frac{1}{3}} = 3^{\frac{1}{3}} = \sqrt[3]{3}$ or $1\cdot44$ (to 3 s.f.)

(iii) $q^{-2} = ({}^{-}4)^{-2} = \frac{1}{16}$ or $0\cdot0625$

(b) (i) $\sqrt[3]{3}$ is irrational.

(ii) Negative whole numbers and numbers which can be expressed as a fraction are rational.

You could also have stated that $\frac{1}{16}$ is a terminating decimal.

(c) $(5 - \sqrt{3})^2 = 25 - 10\sqrt{3} + 3$
$= 28 - 10\sqrt{3}$ or $2(14 - 5\sqrt{3})$

More help or practice

Recurring decimals ► Book YX1 pages 19 to 21
Changing a recurring decimal to a fraction ► Book YX1 page 22
Rational and irrational numbers ► Book YX1 pages 23 to 27
Manipulating surds ► Book YX1 pages 17 to 18

Mixed number (page 18)

1 (a) 13 15 17 19
Sum $= 64 = 4^3$

(b) $\sqrt[3]{729} = 9$; the ninth line.

2 (a) Asia

(b) (i) $1\cdot41 \times 10^8$

(ii) $2\cdot69 \times 10^9 + 0\cdot024 \times 10^9 = 2\cdot714 \times 10^9$

(c) $2\cdot52 \times 10^8 \div 1\cdot93 \times 10^7 = 13\cdot056\ldots$

It would be appropriate to correct your answer to 13 or 13·1, that is, to 2 or 3 s.f.

(d) The lower bound is $510\cdot5 \times 10^6$ and the upper bound is $511\cdot5 \times 10^6$.

Any equivalent form of these answers would gain full marks.

3 The pressure has been increased by a factor of $\frac{1\cdot89}{1\cdot01}$.

Since the volume is inversely proportional to the pressure, the multiplying factor is $\frac{1\cdot01}{1\cdot89}$.

So when the pressure is $1\cdot89 \times 10^5$ Pa, the volume is

$7\cdot3 \times 10^{-4} \times \frac{1\cdot01}{1\cdot89} = 3\cdot9 \times 10^{-4}\,\text{m}^3$ (to 2 s.f.).

Alternatively you could have worked out the constant of proportionality, k, by substituting in $V = \frac{k}{P}$:

$k = 7\cdot3 \times 10^{-4} \times 1\cdot01 \times 10^5$

When P is $1\cdot89 \times 10^5$ Pa,

$$V = \frac{7\cdot3 \times 1\cdot01 \times 10}{1\cdot89 \times 10^5}\,\text{m}^3$$
$$= 3\cdot901\,05\ldots \times 10^{-4}\,\text{m}^3$$
$$= 3\cdot9 \times 10^{-4}\,\text{m}^3 \text{ (to 2 s.f.)}$$

4 (a) $3\frac{3}{4} + 2\frac{1}{2} + 1\frac{1}{4} = 7\frac{1}{2}$

The required fraction is $1\frac{1}{4} \div 7\frac{1}{2} = \frac{5}{4} \div \frac{15}{2}$
$= \frac{5}{4} \times \frac{2}{15}$
$= \frac{1}{6}$

(b) The percentage error $= \frac{1}{4} \div 1\frac{1}{4} \times 100\%$
$= \frac{1}{4} \times \frac{4}{5} \times 100\% = 20\%$

5 (a) (i) $8{\cdot}7 \times 10^9$ (ii) $1{\cdot}87 \times 10^8$
(b) $£(8{\cdot}7 \times 10^9) \div (1{\cdot}87 \times 10^8) = £46{\cdot}52$ (to 2 d.p.)
(c) $\dfrac{8{\cdot}7 \times 10^9}{1{\cdot}87 \times 10^8} \approx \dfrac{9 \times 10^9}{2 \times 10^8} = \frac{90}{2} = 45$
(d) (i) 34 250 and 34 150
(ii) $£4{\cdot}375 \times 10^6$ and $£4{\cdot}365 \times 10^6$
(e) $£42{\cdot}5 \div 1{\cdot}025 = £41{\cdot}46\ldots$
So £41·5 million (to 3 s.f.) was dispensed.

6 (a) 5510
(b) $5510 \times 1{\cdot}08 \times 10^{21}\,\text{kg} = 5{\cdot}9508 \times 10^{24}\,\text{kg}$
(c) $5{\cdot}9508 \times 10^{21}$ tonnes
(d) $(5{\cdot}9508 \times 10^{21}) \div (2 \times 10^{27}) \times 100\%$
$= 3{\cdot}0 \times 10^{-4}\,\%$ (to 2 s.f.)

7 (a) Lower bound $= 132{\cdot}5 \times 6\,\text{g} = 795\,\text{g}$
Upper bound $= 137{\cdot}5 \times 6\,\text{g} = 825\,\text{g}$
(b) It is unlikely that all the bags are at the lower limit.
(c) Maximum possible weight $= 825 \times 1{\cdot}05\,\text{g}$
$= 866({\cdot}25)\,\text{g}$
As the weight could be up to 866 g the label is not misleading.
Minimum possible weight $= 795 \times 0{\cdot}95\,\text{g}$
$= 775({\cdot}25)\,\text{g}$
It is unlikely that the weight would be as great as that quoted, so the label could be misleading.
Either of these answers (or similar) would be acceptable.

ALGEBRA

Simplifying and substituting (page 20)

1 (a) $3ab + a$ (b) $2x - 5y$
(c) $^-18a^2$ (d) a^{-2} or $\dfrac{1}{a^2}$
(e) $10a^3b$ (f) $4cd^3e^2$
(g) $2x + 3y - xy$
Start by multiplying out the brackets.
(h) x^5 (i) $\frac{1}{2}xy$ or $\dfrac{xy}{2}$
(j) $\frac{1}{6}x$ or $\dfrac{x}{6}$ (k) $\dfrac{b^2}{a^2}$ or $a^{-2}b^2$
(l) $\frac{1}{2}a^0 = \frac{1}{2}$ (m) $^-4c^2 - 21c$ or $^-c(4c + 21)$
(n) $3x^2 + 2x^3$ or $x^2(3 + 2x)$
(o) $8m^6$ (p) $9d^5$
Note that letters are usually put in alphabetical order. It's not wrong if you don't do this but it makes it easier to collect like terms.

2 (a) $6(x + 2y)$ (b) $a(3 + b)$ (c) $n(n + 5)$
(d) $5x(2x + y)$ (e) $2x(7 - 3x)$ (f) $x(^-x + 3)$
(g) $^-2x^2(x + 5)$ (h) $\frac{1}{3}\pi r^2(h + 4r)$

3 (a) $^-16\frac{1}{2}$ (b) $^-16$ (c) $^-729$ (d) $^-50$
(e) $^-6\frac{1}{2}$ (f) $\frac{7}{16}$ (g) $\frac{1}{3^3} = \frac{1}{27}$ (h) $^-\frac{3}{5}$
(i) 5 (j) 106 (k) $\frac{1}{125}$ (l) $8^{-\frac{1}{3}} = \frac{1}{2}$

4 $\dfrac{10^6 \times 10^{-3}}{10^{-4}} = 10^3 \times 10^4 = 10^7$

5 (a) 6 (b) $\frac{4}{3}$ or $1\frac{1}{3}$

6 2·5 (to 2 s.f.)

7 $\sqrt{(30 \times 80 \times 0{\cdot}4)} = \sqrt{960} \approx 30$ m.p.h.

8 $V = 649\,\text{cm}^3$ (to 3 s.f.)
Start by evaluating the terms in the brackets.

9 (a) $\pi r(2h + r)$
(b) $1{\cdot}5\pi\left(\frac{2 \times 15}{4} + \frac{3}{2}\right) = 1{\cdot}5\pi\left(\frac{36}{4}\right) = 42{\cdot}411\ldots$
So $A = 42{\cdot}4$ square inches (to 3 s.f.)

10 (a) $e = \frac{4\cdot5}{\sqrt{25}} = 4\cdot5 \div 5 = 0\cdot9$

$Z = \frac{14\cdot5 - 16}{0\cdot9} = {}^{-}1\cdot5 \div 0\cdot9 = {}^{-}1\cdot\dot{6}$ or ${}^{-}1\frac{2}{3}$

(b) $e = 14\cdot37 \div \sqrt{12} = 4\cdot148\ldots$

$Z = (7\cdot81 + 3\cdot64) \div 4\cdot148\ldots$

$= 2\cdot76$ (to 3 s.f.)

Remember to round only at the end of a calculation: using a rounded value for e would lose marks. Also, always write down the value of any intermediate step which could score marks.

More help or practice

Substituting values in expressions ► Book Y1 pages 21 to 30, 63

Simplifying by removing single brackets
► Book Y1 pages 104 to 111, Book Y2 pages 45 to 47

Simplifying by collecting terms ► Book Y2 pages 43 to 44

Simplifying by factorising (single brackets only)
► Book Y2 page 49

The rules of indices ► Book YR+ pages 29 to 30, Book YX2 pages 27 to 28

Manipulating fractional indices ► Book YX2 pages 28 to 31

Using indices ► Book YX2 pages 36 to 37

Solving linear equations (page 22)

After solving an equation it is important to check your answer by substituting your solution back into the original equation.

1 (a) $x = 4$ (b) $a = \frac{2}{3}$ (c) $x = 10$

(d) $x = 14$ *Start by multiplying by 3.*

(e) $w = 0\cdot6$

(f) $2x = 9$, so $x = 4\frac{1}{2}$ or $4\cdot5$

Remember to collect terms in x on one side.

(g) $x = {}^{-}2\cdot2$

(h) ${}^{-}\frac{5}{2}x = {}^{-}3$, so $x = \frac{6}{5}$ or $1\frac{1}{5}$ or $1\cdot2$

(i) $4x - 6 = {}^{-}10$, so $x = {}^{-}1$

Multiply by 5 and then multiply out the brackets.

(j) $2x - 5 = 21 - 3x$, so $5x = 26$ and $x = 5\cdot2$

(k) $5x - 17\cdot5 = 8$, so $5x = 25\cdot5$ and $x = 5\cdot1$

(l) $x = 0$

(m) $k = 1\cdot47$

(n) $10z - 3z = 70$, so $z = 10$

(o) $2x - 18 = 9 - 3x$, so $5x = 27$ and $x = 5\cdot4$

(p) $16 - 10 + 4x = 1$, so $4x = {}^{-}5$ and $x = {}^{-}1\cdot25$

2 (a) (i) $3x$ (ii) $x - 9$ (iii) $3x - 9$

(b) (i) $3x - 9 = 6(x - 9)$ which simplifies to $3x = 45$.

Any correct form of the equation should gain full marks but you will need to simplify it to solve it.

(ii) $x = 15$

3 $5(x + 3) = 16$ which simplifies to $5x = 1$.
So $x = 0\cdot2$ and the length of the unknown side is $0\cdot2 + 3 = 3\cdot2$.

More help or practice

Solving linear equations ► Book Y1 pages 64 to 70

Looking at graphs (page 23)

1 (a) $3y = {}^{-}2x + 12$

$y = {}^{-}\frac{2}{3}x + 4$

(b) $y = {}^{-}\frac{2}{3}x$ or $2x + 3y = 0$

Lines through the origin have no constant term.

2 (a) $y = 2x^2 + 3$ (b) $y = \frac{2}{x}$

(c) $y = 3 - 2x^2$ (d) $y = 2x^3$

More help or practice

Gradients and intercepts of straight lines
► Book Y3 pages 11 to 22, Book Y5 pages 70 to 72

The graphs of $y = ax^2$, $y = ax^3$ and $y = \frac{a}{x}$
► Book Y5 pages 73 to 75

Simultaneous equations (page 24)

1 $x = \frac{1}{2}, y = 4\frac{1}{2}$

*Remember that you only **need** to plot two points to draw a straight line, but it is a good idea to check with a third. Always look for 'easy' points such as where the line cuts the x- or y- axis. In this case the easiest points are where $x = 0$ (giving $y = 5$ in the first equation and $y = 3$ in the second) and $y = 0$ (giving $x = 5$ in the first equation and $x = {}^{-}1$ in the second). Then choose a third point to check.*

2 $x = 0{\cdot}5$ and $y = 2{\cdot}7$

3 (a) $x = 7, y = 3$ (b) $x = 2, y = 4$
(c) $x = 3, y = {}^{-}2$ (d) $x = 6, y = 5$

4 *Always be careful with units when forming equations. In this question there are two possible equivalent equations:*

(a) $(\frac{1}{2} \times x) + (1 \times y) = 140$ and $50x + 100y = 14\,000$
Simplifying either of these equations gives $x + 2y = 280$.

(b) $x + y = 229$

(c) Subtracting the equations gives $y = 51$.
Substituting gives $x = 178$.
So 178 articles were sold at 50p and 51 at £1.
Check by substituting $x = 178$ and $y = 51$ in one of the equations: $178 + 51 = 229$.

5 The two simultaneous equations $2a + b = 34$ and $a + 2b = 32$ solve to give $a = 12$ and $b = 10$.

6 Let x years be the age of the mother and y years the age of the daughter.
Then $x + y = 43$
and $x - 4 = 6(y - 4)$ which simplifies to $x - 6y = {}^{-}20$.
Solving the equations gives $y = 9$ and $x = 34$.
The mother is 34 years old and the daughter is 9 years old.

7 (a) $x = 2$ and $y = 3$

(b) (i) Comparing equations gives
$p^{-1} = 2$ and $q^{\frac{1}{2}} = 3$.

(ii) $\frac{1}{p} = 2$ and $\sqrt{q} = 3$ which solve to give
$p = \frac{1}{2}$ and $q = 9$.

More help or practice
Graphical solution ► Book Y2 pages 130 to 132
Algebraic solution ► Book Y3 pages 135 to 139, Book Y4 pages 20 to 24

Inequalities and regions (page 26)

1 (a) $x > {}^{-}2$

(b) *You need to collect terms in x on one side of the equation. Here are two alternative methods.*

$$x + 7 < 3x + 2 \quad \text{or} \quad 3x + 2 > x + 7$$
$$^{-}2x < {}^{-}5 \qquad\qquad 2x > 5$$
$$x > \frac{5}{2} \qquad\qquad x > \frac{5}{2}$$

(c) $y > {}^{-}4$ (d) $n \geq {}^{-}4$ (e) $x > {}^{-}2$
(f) $x < {}^{-}2$ (g) $a \leq 3\frac{1}{3}$ (h) $x < {}^{-}2$ or $x > 2$

2 The unshaded region satisfies all the inequalities, including the lines forming the sides of the triangle.

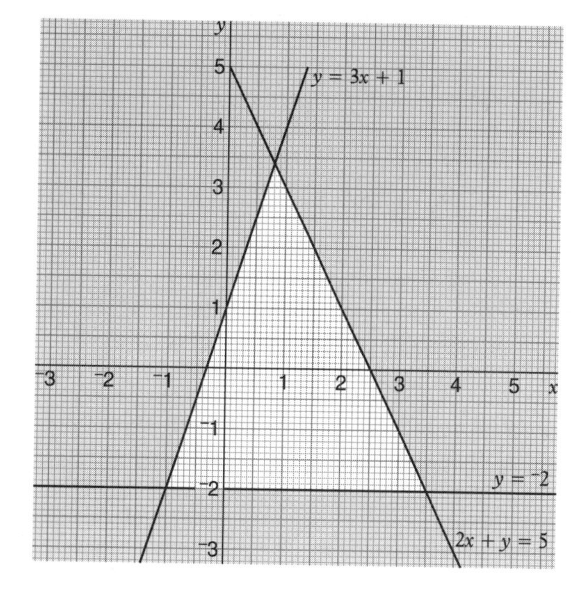

3 (a) Start by solving the two inequalities
$3x + 1 \leq 27$ and $5x - 6 \geq 27$.
So $x \leq 8\frac{2}{3}$ and $x \geq 6\frac{3}{5}$.

(i) The largest value of x is $8\frac{2}{3}$.

(ii) The smallest value of x is $6\frac{3}{5}$.

(b) The integer values of x are 7 and 8.

4 (a) Solving the inequality $13 - 6x < 25$ gives $x > {}^{-}2$.

(b) $y = 19$

y has its maximum value when z is a maximum (that is 7) and $y - z = 12$.

5 (a) $y < x + 1$ or $y \leq x + 1$ and $y > x^2$ or $y \geq x^2$

(b) At P both $y = x + 1$ and $y = x^2$, so $x^2 = x + 1$.

6 (a) False (b) True (c) False

(d) False (e) True

7 (a) $y \geq 10$ or $y > 9$

(b) $x + y \leq 30$ or $x + y < 31$

(c)

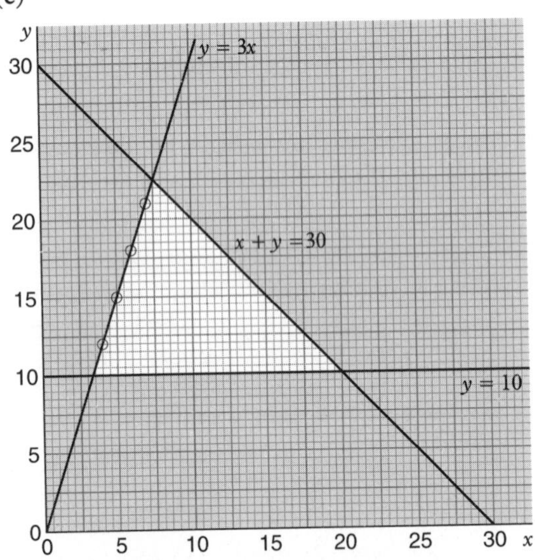

The unshaded region satisfies both inequalities.

(d) The line with equation $y = 3x$; see the graph.

(e) There are four possible answers: (4, 12), (5, 15), (6, 18) and (7, 21).

More help or practice

Solving inequalities ► Book Y5 pages 92 to 97

Inequalities and regions ► Book Y5 pages 121 to 124

Regions with two boundaries ► Book Y5 pages 125 to 126

Trial and improvement; graphical solution (page 28)

1 *If you show your trials you may get some marks for them even if your final answer is wrong.*
Here are some trials. You may have needed a few more.

Trial values of x	Value of $x^3 - 5x^2 + x - 10$
5	⁻5
6	32
5·5	10·625
5·2	0·608
5·18	0·009832
5·17	⁻0·286087
5·175	⁻0·138391

The solution is 5·18 (to 3 s.f.).

2 *You need to find an approximate solution for your first trial value.*

$2x^3 = 15$ so $x^3 = 7{\cdot}5$ and x is slightly less than 2.

Trial values of x	$2x^3$
1·9	13·718
1·95	14·82975
1·96	15·059072
1·955	14·94411775

The solution must lie between 1·955 and 1·96, which is 1·96 (to 2 d.p.).

3 (a)

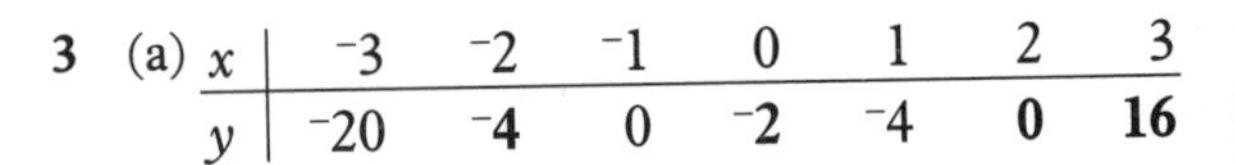

x	⁻3	⁻2	⁻1	0	1	2	3
y	⁻20	**⁻4**	0	**⁻2**	⁻4	**0**	**16**

(b)

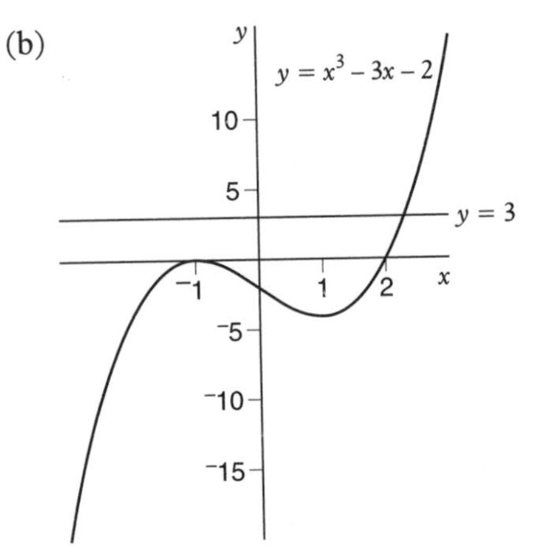

(c) Draw the line $y = 3$; its point of intersection with the curve is $x = 2{\cdot}3$.

(d) Rewrite the equation as $x^3 - 3x - 27 = 0$.
From the graph we can see that the solution is likely to lie between $x = 3$ and $x = 4$.

Trial values of x	Value of $x^3 - 3x - 27$
3	$^-9$
3·5	5·375
3·3	$^-0·963$
3·4	2·104
3·35	0·545375

The solution must lie between 3·3 and 3·35, which is 3·3 (to 1 d.p.).

4 (a)

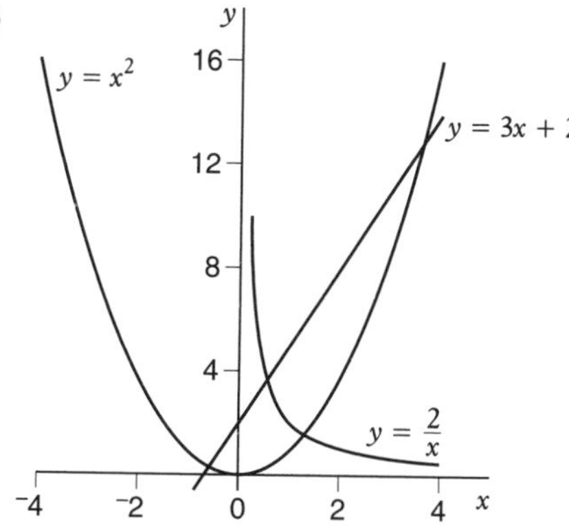

(b) The solutions are $x = 3·6$ and $x = {}^-0·6$.
These are the x-coordinates of the points where $y = x^2$ cuts the line $y = 3x + 2$.

(c) (i) You need to draw the graph of $y = \frac{2}{x}$ for positive values of x.

First draw up a table of values of x.

x	1	2	3	4	$\frac{1}{2}$	$\frac{1}{4}$
y	2	1	$\frac{2}{3}$	$\frac{1}{2}$	4	8

Then plot the points and draw the curve. Notice how you need to add at least one fractional value of x.

(ii) $x = 0·5$ or $0·6$

(d) $y = {}^-4x + 1$
Rewrite the equation $x^2 + 4x - 1 = 0$ in the form $x^2 = {}^-4x + 1$.

5 (a)

x	0·5	1	2	3	4	5	6
$\frac{x^2}{10}$	0·025	0·1	0·4	0·9	1·6	2·5	3·6
$\frac{^-1}{x}$	$^-2$	$^-1$	$^-0·5$	$^-0·\dot{3}$	$^-0·25$	$^-0·2$	$^-0·1\dot{6}$
y	$^-1·975$	$^-0·9$	$^-0·1$	$0·5\dot{6}$	1·35	2·3	$3·4\dot{3}$

(b)

y
4
2
0
−2
2 4 6 x
$y = \frac{x^2}{10} - \frac{1}{x}$
$y = 3 - 2x$

(c) (i) The point where the curve cuts the x-axis is the solution of $y = \frac{x^2}{10} - \frac{1}{x} = 0$ or $x^3 - 10 = 0$.
This is $\sqrt[3]{10}$.

(ii) Your calculator will give $\sqrt[3]{10} = 2·154\ldots$ but an answer of 2·1 or 2·2 will gain full marks.

(d) See the graph.

(e) 1·6 or 1·7

(f) At the point where the two graphs intersect,
$\frac{x^2}{10} - \frac{1}{x} = 3 - 2x$.
Multiplying throughout by $10x$ gives
$x^3 - 10 = 30x - 20x^2$ or
$x^3 + 20x^2 - 30x - 10 = 0$.
The x-coordinate of the point of intersection gives an approximate solution to this equation.

More help or practice
Trial and improvement ► Book Y4 pages 149 to 152
Drawing graphs of quadratic equations ► Book Y4 pages 143 to 145
Graphical solution of polynomials ► Book YX1 pages 55 to 57

Gradients and tangents (page 30)

1 (a) About 8

Any answer in the range 7 to 9 would gain full marks.

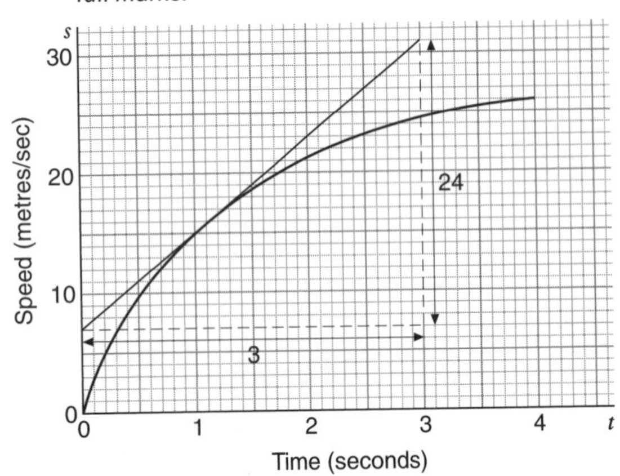

(b) The gradient represents an acceleration of 8 m/s^2 (or 8 m s^{-2}).

2 The rate of cooling is 9°C/minute.

Draw a tangent at $t = 2$ and estimate the gradient of the line to find this.

An answer of 8 °C or 10 °C/minute would also be acceptable.

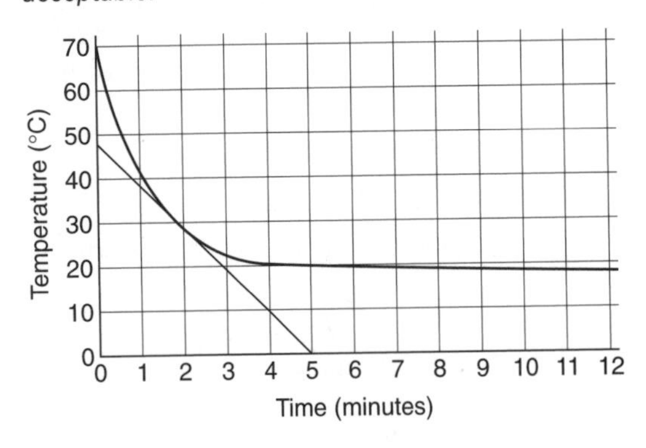

3 (a) (i) 0·52 seconds

(ii) The average speed $= \frac{2\cdot0}{0\cdot52}$ m/s

$= 3\cdot8$ m/s (to 1 d.p.)

(b) Any answer between 5 m/s and 7 m/s

Remember that it makes no sense in this context to give an answer to more than 1 or 2 significant figures.

(c) (i) $1 = 2 - 5t^2(1 + t)$

(ii) $5t^2(1 + t) = 1$

$$t^2 = \frac{1}{5(1+t)}, \text{ so } t = \frac{1}{\sqrt{(5(1+t))}}$$

More help or practice

Gradients and straight lines ► Book YX1 pages 28 to 29

Tangents to a curve ► Book YX1 pages 30 to 37

Quadratics (page 31)

1 (a) $2x^2 + 13x - 7$ (b) $9n^2 - 1$

(c) $9x^2 - 12x + 4$ (d) $^-20a^2 + 7a + 49$

2 (a) $x(x + 4) = 0$, so $x = 0$ or $x = {}^-4$

(b) $(x + 3)(x + 4) = 0$, so $x = {}^-3$ or $x = {}^-4$

(c) $(x - 7)(x + 2) = 0$, so $x = 7$ or $x = {}^-2$

(d) *First write the equation in the form* $x^2 - x - 12 = 0$.
$(x - 4)(x + 3) = 0$, so $x = 4$ or $x = {}^-3$

(e) $(x + 10)(x - 8) = 0$, so $x = {}^-10$ or $x = 8$

(f) *Rearrange the equation as* $x^2 - 3x - 54 = 0$.
$(x - 9)(x + 6) = 0$, so $x = 9$ or $x = {}^-6$

(g) $x^2 + 3x - 28 = 0$
$(x + 7)(x - 4) = 0$, so $x = {}^-7$ or $x = 4$

(h) *Multiply out the brackets and collect terms on one side of the equation.*
$6x^2 + 3x = 12 - 3x$
$6x^2 + 6x - 12 = 0$
$6(x^2 + x - 2) = 0$ *Take out the factor 6.*
$6(x + 2)(x - 1) = 0$, so $x = {}^-2$ or $x = 1$

(i) $x^2 + 6x + 9 = 2x + 12 + 9$
$x^2 + 4x - 12 = 0$
$(x + 6)(x - 2) = 0$, so $x = {}^-6$ or $x = 2$

3 (a) *Substitute $a = 1$, $b = {}^-15$ and $c = 50$ into the standard formula* $\frac{^-b \pm \sqrt{(b^2 - 4ac)}}{2a}$.

$$x = \frac{15 \pm \sqrt{(225 - 200)}}{2} = \tfrac{1}{2}(15 \pm \sqrt{25})$$

$$= \tfrac{1}{2}(15 + 5) \text{ or } \tfrac{1}{2}(15 - 5)$$

$$= 10 \text{ or } 5$$

(b) *Substitute $a = 1$, $b = 20$ and $c = {}^-32$ into the standard formula.*

$$\frac{^-20 \pm \sqrt{(400 + 128)}}{2} = \tfrac{1}{2}(^-20 \pm \sqrt{528})$$

$$= {}^-21\cdot49 \text{ or } 1\cdot49 \text{ (to 2 d.p.)}$$

(c) $x = \frac{1}{2}(^-15 \pm \sqrt{(225 + 200)}) = \frac{1}{2}(^-15 \pm \sqrt{425})$
$= {^-17{\cdot}81}$ or $2{\cdot}81$ (to 2 d.p.)

(d) $x = \dfrac{^-12 \pm \sqrt{(144 + 680)}}{4} = \frac{1}{4}(^-12 \pm \sqrt{824})$
$= {^-10{\cdot}18}$ or $4{\cdot}18$ (to 2 d.p.)

(e) *First re-write the equation in standard form.*
$2x^2 - 5x - 2 = 0$
$x = \dfrac{5 \pm \sqrt{(25 + 16)}}{4} = \frac{1}{4}(5 \pm \sqrt{41})$
$= 2{\cdot}85$ or ${^-0{\cdot}35}$ (to 2 d.p.)

(f) $3x^2 - 8x - 8 = 0$
$x = \dfrac{8 \pm \sqrt{(64 + 96)}}{6} = \frac{1}{6}(8 \pm \sqrt{160})$
$= 3{\cdot}44$ or ${^-0{\cdot}77}$ (to 2 d.p.)

More help or practice
Expanding double brackets ► Book Y2 page 48, Book Y4 page 145
Factorising quadratic equations ► Book Y4 pages 147 to 148

Algebraic fractions (page 32)

1 (a) $\dfrac{x + y}{xy}$

(b) $\dfrac{2(x + 1) + x + 4}{(x + 4)(x + 1)} = \dfrac{3x + 6}{(x + 4)(x + 1)}$
$= \dfrac{2(x + 3)}{(x + 4)(x + 1)}$

(c) $\dfrac{3(x + 3) - 2(x - 4)}{(x - 4)(x + 3)} = \dfrac{x + 17}{(x - 4)(x + 3)}$

(d) $\dfrac{4x(x + 7) - (x - 5)(x - 2)}{(x - 2)(x + 7)} = \dfrac{3x^2 + 35x - 10}{(x - 2)(x + 7)}$

(e) $\dfrac{3x^2 + y^2}{yx}$

2 (a) $\dfrac{30 + 9(1 - 2z)}{10(1 - 2z)} = \dfrac{3}{2}$
$39 - 18z = 3 \times 5(1 - 2z)$
$39 - 18z = 15 - 30z$
$12z = {^-24}$
$z = {^-2}$

(b) $\dfrac{2(x + 3) + 5(x - 3)}{(x - 3)(x + 3)} = 1$
$2x + 6 + 5x - 15 = x^2 - 9$
$x^2 - 7x = 0$
$x(x - 7) = 0$
$x = 0$ or $x = 7$

(c) $\dfrac{x(x - 2) + 2x(x + 1)}{(x + 1)(x - 2)} = 3$
$x^2 - 2x + 2x^2 + 2x = 3(x^2 - x - 2)$
$3x^2 = 3x^2 - 3x - 6$
$0 = {^-3x} - 6$
$x = {^-2}$

(d) $\dfrac{(y + 1)(y + 8) - 2y(y - 1)}{(y - 1)(y + 8)} = 1$
$y^2 + 9y + 8 - 2y^2 + 2y = y^2 + 7y - 8$
${^-2y^2} + 4y + 16 = 0$
$y^2 - 2y - 8 = 0$
$(y - 4)(y + 2) = 0$
$y = 4$ or $y = {^-2}$

Remember to check your answers by substituting in the original equation.

3 (a) $\dfrac{2}{x - 1} - \dfrac{3}{x + 2} = 4$
$\dfrac{2(x + 2) - 3(x - 1)}{(x - 1)(x + 2)} = 4$
$2x + 4 - 3x + 3 = 4(x^2 + x - 2)$
$7 - x = 4x^2 + 4x - 8$
$0 = 4x^2 + 4x - 8 - 7 + x$
So $4x^2 + 5x - 15 = 0$

(b) $x = 1{\cdot}41$ or ${^-2{\cdot}66}$ (to 2 d.p.)

More help or practice
Algebraic fractions ► Book Y4 pages 33 to 39, Book Y5 pages 16 to 20

Changing the subject of a formula (page 33)

1 (a) $3V^2T = K$ so $T = \dfrac{K}{3V^2}$

(b) $\sqrt{a} = \dfrac{b}{2k}$ so $a = \dfrac{b^2}{4k^2}$

(c) $2as = v^2 - u^2$ so $a = \dfrac{v^2 - u^2}{2s}$

(d) $v^2 - u^2 = 2as$

$v^2 = 2as + u^2$

$v = \pm\sqrt{(u^2 + 2as)}$

(e) $mp = \sqrt{(n^2 - p^2)}$

$m^2p^2 = n^2 - p^2$

$m^2p^2 + p^2 = n^2$

$p^2(m^2 + 1) = n^2$

$p = \pm\sqrt{\left(\dfrac{n^2}{m^2+1}\right)} = \dfrac{n}{\pm\sqrt{(m^2+1)}}$

(f) $x^2t^2 = k(t+1)$

$x = \dfrac{\pm\sqrt{(k(t+1)}}{t}$

(g) $k(r+h) = rh$

$rh - kh = kr$

$h(r-k) = kr$

$h = \dfrac{kr}{r-k}$

(h) $\dfrac{1}{u} = \dfrac{1}{f} - \dfrac{1}{v} = \dfrac{v-f}{fv}$

$u = \dfrac{fv}{v-f}$

2 (a) $m(v-u) = Ft$ so $m = \dfrac{Ft}{v-u}$

(b) $m = \dfrac{52{\cdot}00 \times 0{\cdot}76}{10{\cdot}40} = 3{\cdot}8\,\text{kg}$

3 (a) $\dfrac{4}{3\frac{1}{4}} + 6 = (4 \times \frac{4}{13}) + 6 = 1\frac{3}{13} + 6 = 7\frac{3}{13}$

(b) *In this question it is probably best to make x the subject of the equation and then substitute the value for y. Start by multiplying each term of the equation by 7 – x to remove the fraction.*

$y(7-x) = 4 + 6(7-x)$

$7y - yx = 4 + 42 - 6x$

$x(6-y) = 46 - 7y$

$x = \dfrac{46 - 7y}{6 - y}$

$= \dfrac{46 + 21}{6 + 3} = \dfrac{67}{9} = 7\frac{4}{9}$

4 (a) $r = \dfrac{2{\cdot}01 \times 10^{-23} \times 2{\cdot}0 \times 10^7}{2 \times 1{\cdot}6 \times 10^{-20} \times 2000} = \dfrac{4{\cdot}02 \times 10^{-16}}{6{\cdot}4 \times 10^{-17}}$

$= 6{\cdot}281\ldots = 6{\cdot}3$ (to 1 d.p.)

(b) $v = \dfrac{2erH}{M}$

5 *First square both sides of the equation, then multiply both sides by x, before gathering terms on the left.*

$x^2 = 3 + \dfrac{1}{x}$

$x^3 = 3x + 1$

$x^3 - 3x - 1 = 0$

Comparing terms with $x^3 + px^2 + qx + r = 0$ gives $p = 0, q = {}^-3$ and $r = {}^-1$.

More help or practice

Manipulating formulas ► Book Y2 pages 77 to 79, 94 to 97; Book Y4 pages 68 to 74

Using algebra (page 34)

1 (a) 3, 7, 11, 15, 19

(b) 4, 7, 12, 19, 28

(c) 2, 3, 5, 9, 17 *Remember that $2^0 = 1$.*

(d) 0, 2, 6, 12, 20

2 (a) $3n$ (b) 3^n (c) $3 \times 2^{n-1}$ (d) $2n^2$

3 $5x^2 + 3x$ or $x(5x + 3)$

4 Let the number be n.

Then $2n + 18 = (n-3)^2$

$2n + 18 = n^2 - 6n + 9$

$n^2 - 8n - 9 = 0$

$(n-9)(n+1) = 0$

$n = 9$

The solution $n = {}^-1$ is impossible because n is positive.

5 (a) Aaron is $x + 3$ years old.
Jason is $x - 2$ years old.
So $(x + 3)(x - 2) = 126$ and
$x^2 + x - 132 = 0$

(b) *Solve the quadratic equation to find x and hence the ages of all three children.*
$(x + 12)(x - 11) = 0$
$x = {}^-12$ or 11
So Maria is 11, Aaron is 14 and Jason is 9 years old.

6 (a) *In questions like this it usually helps to first find the next couple of terms to see how the pattern is built up. In this case we have rectangle numbers; the 4th term is 5 × 6 = 30 and the 5th term is 6 × 7 = 42.*
The number of counters in the nth pattern is $(n + 1)(n + 2) = n^2 + 3n + 2$.
Either form of the expression would gain full marks but if you wrote $n + 1 \times n + 2$ you would lose a mark.

(b) The sequence is 6, 12, 20, 30, 42, ...
The differences are 6, 8, 10, 12, ...
The nth difference (which added to the nth term will give the $(n + 1)$th term) is $2(n + 2)$ or $2n + 4$.
So Glyn will need an extra $2n + 4$ counters.

7 (a) $x(30 - 2x)$

(b) $x(30 - 2x) = 100$
$30x - 2x^2 = 100$
$x^2 - 15x + 50 = 0$
$(x - 5)(x - 10) = 0$
$x = 5$ or 10

(c) Both solutions are valid but they give different shapes ($x = 5$ gives a 5 m by 20 m rectangle and $x = 10$ gives a square of side 10 m).

8 $(n - 1)^2 + n^2 + (n + 1)^2 = 365$
$n^2 - 2n + 1 + n^2 + n^2 + 2n + 1 = 365$
$3n^2 = 363$
$n^2 = 121$
$n = \pm 11$
The three positive numbers are 10, 11 and 12.

9 $n(3n + 1) = 1344$
$3n^2 + n - 1344 = 0$
Using the quadratic formula,
$$n = \frac{{}^-1 \pm \sqrt{(1 + 16\,128)}}{6}$$
$$= \frac{{}^-1 \pm 127}{6}$$
$$= 21 \text{ or } {}^-21\tfrac{1}{3}$$
21 terms add up to 1344.

10 (a) The length of the second side of the rectangle is $(8 - a)$ cm.
So $a(8 - a) = 10$
$8a - a^2 = 10$
$a^2 - 8a + 10 = 0$

(b) $$a = \frac{8 \pm \sqrt{(64 - 40)}}{2}$$
$= 6{\cdot}45$ or $1{\cdot}55$ (to 2 d.p.)

11 (a) The product of the inner pair is always two more than the product of the outer pair.

(b) Let the numbers be $n, n + 1, n + 2$ and $n + 3$.
The difference betwen the two respective products is
$(n + 1)(n + 2) - n(n + 3)$
$= n^2 + 3n + 2 - n^2 - 3n$
$= 2$
So the rule found in (a) always works.

(c) Let the numbers be $n, n + 2, n + 4$ and $n + 6$, where n is odd.
$(n + 2)(n + 4) - n(n + 6)$
$= n^2 + 6n + 8 - n^2 - 6n$
$= 8$
So, for consecutive odd numbers, the difference between the two products is 8.
Note that the difference would be the same for consecutive even numbers.

More help or practice

Finding rules for sequences ► Book Y4 pages 53 to 61
Pattern spotting ► Book YR+ pages 44 to 46

Functions and graphs (page 36)

1 (a) $f(5) = 16 + 4 = 20$ (b) $f(1) = 0 + 4 = 4$

(c) $f(a + 1) = a^2 + 4$

(d) $f(^-3) = 16 + 4 = 20$

(e) $f(\frac{1}{2}) = \frac{1}{4} + 4 = 4\frac{1}{4}$

2 (a)
$$2(x + 1) + 3 = 4 - 3(x - 1)$$
$$2x + 2 + 3 = 4 - 3x + 3$$
$$5x = 2$$
$$x = \frac{2}{5}$$

(b)
$$2(2x) + 3 = 3(4 - 3x)$$
$$4x + 3 = 12 - 9x$$
$$13x = 9$$
$$x = \frac{9}{13}$$

(c)
$$\frac{1}{2}(2x + 3) = 4 - 3(x + 5)$$
$$x + \frac{3}{2} = 4 - 3x - 15$$
$$4x = \frac{^-25}{2}$$
$$x = \frac{^-25}{8} = {^-3}\frac{1}{8}$$

Check your solutions by substituting in both sides of the original equation.

3

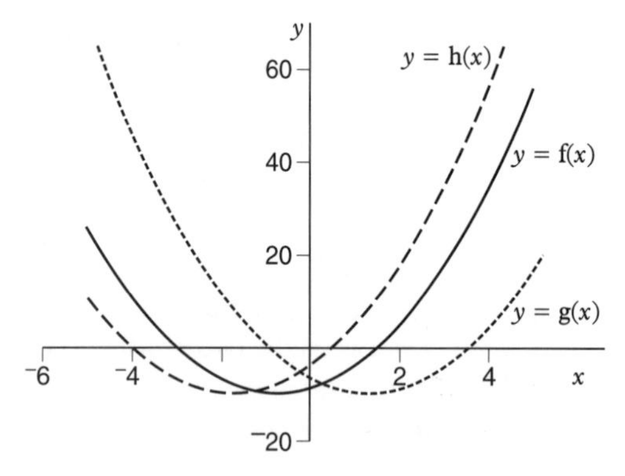

$g(x) = f(x - 2) = 2x^2 - 5x - 7$

Remember that if the graph of $f(x)$ *is translated by* k *units to the* ***right****, it becomes the graph of the function* $\mathbf{f(x - k)}$. *It is easy to put* $f(x + k)$ *by mistake.*

$h(x) = f(x + 1) = 2x^2 + 7x - 4$

The solutions of $2x^2 + 3x - 9 = 0$ are $^-3$ and $1\frac{1}{2}$,

the solutions of $2x^2 - 5x - 7 = 0$ are $^-1$ and $3\frac{1}{2}$, and

the solutions of $2x^2 + 7x - 4 = 0$ are $^-4$ and $\frac{1}{2}$.

From these we can see that the graphs cut the x-axis at the expected points.

4 The two points $x = 2$ and $x = 6$ are translated to $x = {^-5}$ and $x = {^-1}$. This would indicate a translation of 7 units to the left.

Confirm this by working out
$$f(x + 7) = {^-(x+7)^2} + 8(x+7) - 12$$
$$= {^-x^2} - 6x - 5$$

5 For $g(x)$ $a = 0$, $b = 3$

For $h(x)$ $a = {^-4}$, $b = 0$

For $m(x)$ $a = {^-4}$, $b = 3$

6 (a) $y = \frac{1}{2}f(x)$ (b) $y = 1 + f(x)$

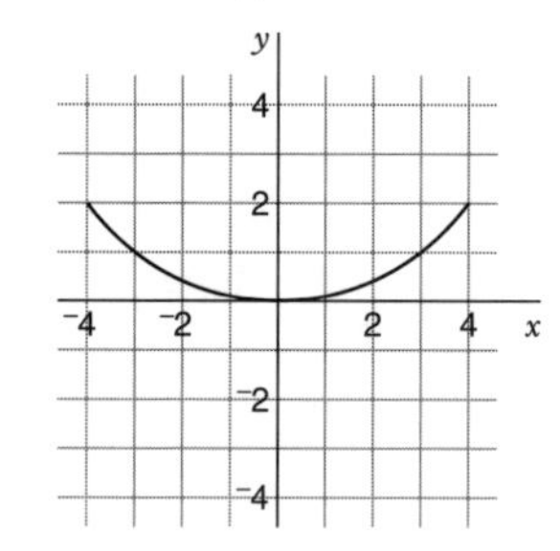

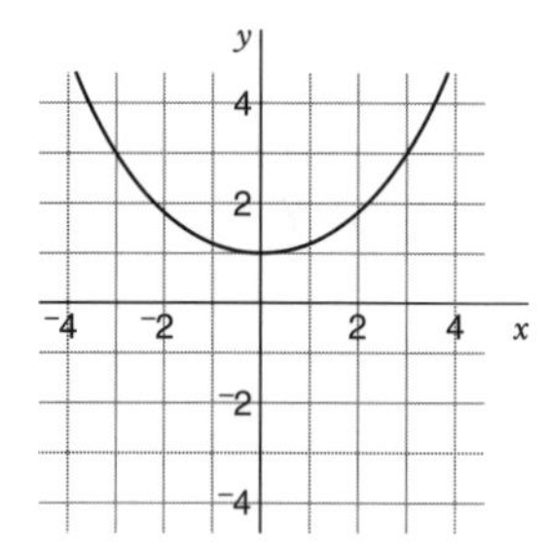

(c) $y = f(\frac{x}{2})$

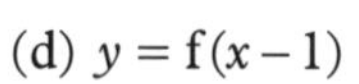
(d) $y = f(x - 1)$

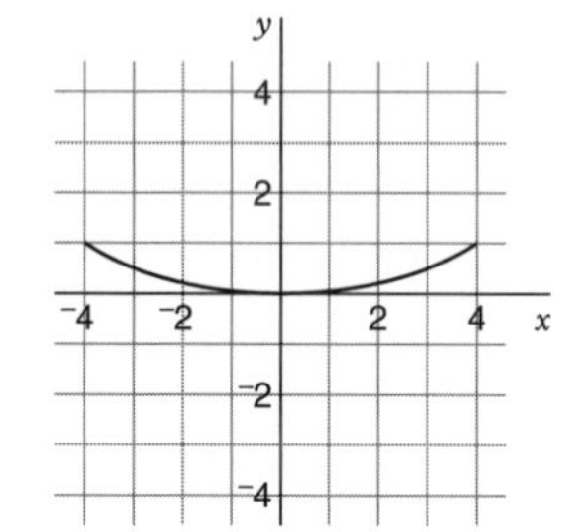

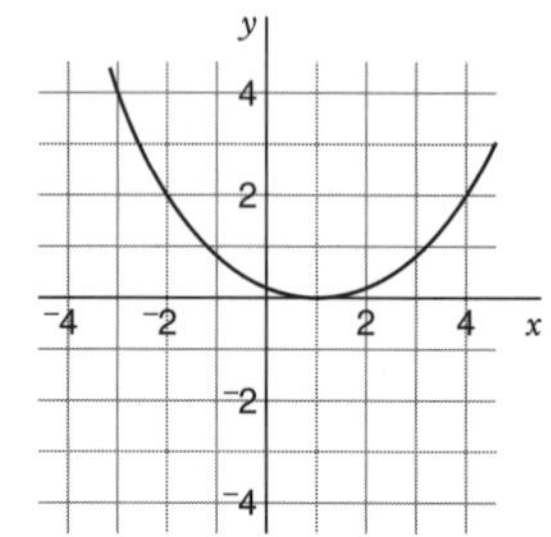

More help or practice

Functions ► Book Y4 pages 122 to 125

Functions and graphs ► Book YX2 pages 65 to 67

The area under a graph (page 38)

1 (a) 60 joules (b) $42\frac{1}{2}$ joules

2 (a) $y = 0, 0{\cdot}259, 0{\cdot}500, 0{\cdot}707, 0{\cdot}866, 0{\cdot}966, 1$

(b) Area = 49·5 square units (using the lower rectangles and the first six values of y).
Any answer between 49 and 50 is acceptable.

(c) Area = 64·5 square units (using the upper rectangles and the second to the last values of y).
Any answer between 64 and 65 is acceptable.

(d) The mean area is 57 square units (to 2 s.f.).
Your answer will depend on your answers to (b) and (c) but should be between 56·5 and 57·5 square units.

(e) $\frac{1}{2} \times 15 \times (0 + 2\times 0{\cdot}259 + 2\times 0{\cdot}5 + 2\times 0{\cdot}707 + 2\times 0{\cdot}866 + 2\times 0{\cdot}966 + 1)$
$= 7{\cdot}5 \times 7{\cdot}596$
$= 56{\cdot}97$
So area = 57 square units (to 2 s.f.)

3 1660 square centimetres (using a width of 20 units)
1740 square centimetres (using a width of 10 units)

4 (a) The measurements are approximately those of a statue on Easter Island.

(b) 37 cubic metres

(c) 110 tonnes (to 2 s.f.)

More help or practice
The area under a graph ► Book Y5 pages 31 to 37, 41 to 43
The trapezium rule ► Book Y5 pages 38 to 40

Fitting functions to data (page 40)

1

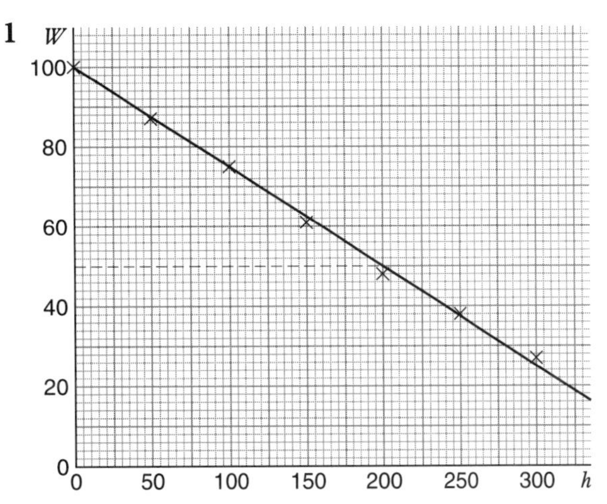

The gradient of the line is given by $^-50 \div 200 = {^-0{\cdot}25}$ and the intercept on the W-axis is 100, so the equation of the line is $W = {^-0{\cdot}25}h + 100$.

2 (a) matches (ii) (b) matches (i)
(c) matches (iii) (d) matches (i)
Note that only even powers of x have the same y-values for positive and negative values of x.

3 (a) The values of D^2 are 0·36, 0·81, 2·25 and 4·00.

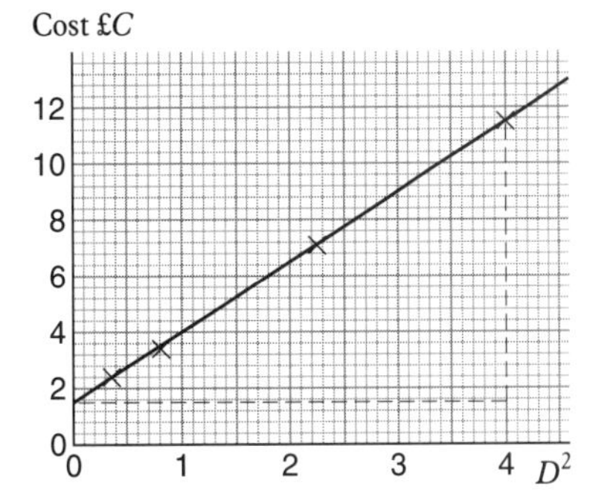

(b) (i) Yes, because the graph of C against D^2 is a straight line.

(ii) $p \approx 2\frac{1}{2}$
The gradient of the line is given by 10 ÷ 4 (see the graph above).
$q \approx 1\frac{1}{2}$ *This is the intercept on the C-axis.*

4 (a) $y = px^3 + q$ (b) $q = 2$ (c) $p = 0{\cdot}25$

To find the value of p substitute a pair of known values of x and y, for example (2, 4), in $y = px^3 + 2$.
Check your answer with another pair, for example ($^-2$, 0).

5 The values of $\frac{1}{M}$ are 1·25, 0·83, 0·67, 0·56, 0·53 and 0·45.

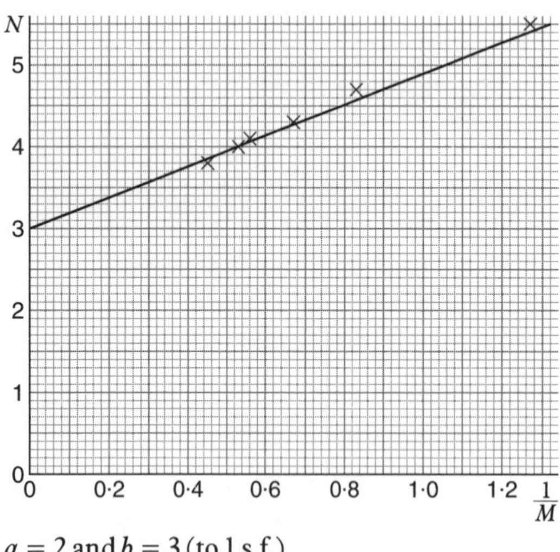

$a = 2$ and $b = 3$ (to 1 s.f.)

6 *Start by working out a table of values for $\frac{1}{A}$ and $\frac{1}{A^2}$.*

A	4	5·6	8	11	16	22
S	60	30	15	8	4	2
$\frac{1}{A}$	0·25	0·179	0·125	0·091	0·063	0·045
$\frac{1}{A^2}$	0·063	0·032	0·016	0·008	0·004	0·002

If you start to plot S against $\frac{1}{A}$ it is immediately obvious that there is not a proportional relationship because the line does not pass through the origin.
However if you plot S against $\frac{1}{A^2}$ the points lie on an approximate straight line through the origin, so Janine is correct.
The line of best fit has gradient 970, so the equation connecting S and A is $S = \frac{970}{A^2}$.

7 (a) $k = 2$ and $A = 4$

(b) *Substitute $t = 0$ and $P = 2$ to find the value of a.*
$2 = ak^0 = a$ $(k^0 = 1)$
Then substitute, say, $P = 0{\cdot}4$ and $t = 10$ to find k.
$0{\cdot}4 = 2k^{10}$ and $k = \sqrt[10]{0{\cdot}2} = 0{\cdot}85$ (to 2 s.f.)

8 (a) First try plotting L against d^2. The line of best fit has gradient 16, so $L = 16d^2$.

(b) $P = 0{\cdot}15d^2 + 7$

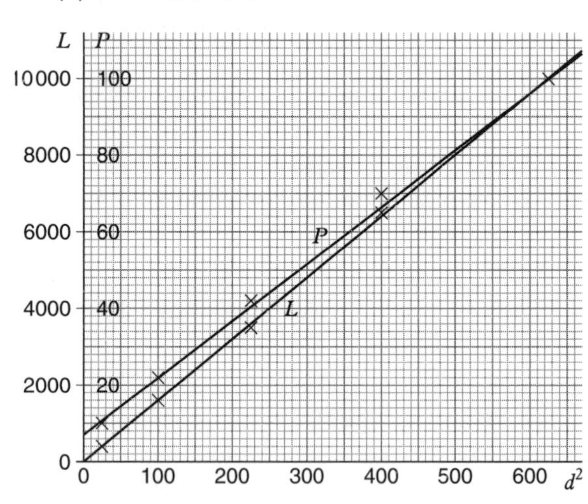

More help or practice

Gradient and intercept of a straight line
► Book Y5 pages 70 to 72
The graphs of $y = ax^2$, $y = ax^3$ and $y = \frac{a}{x}$
► Book Y5 pages 73 to 75
Fitting a linear equation (line of best fit)
► Book Y5 pages 76 to 77
Fitting non-linear equations ► Book Y5 pages 77 to 79, Book YX1 pages 74 to 78
Exponential graphs ► Book YX2 pages 33 to 35

Mixed algebra (page 43)

1 (a) (i) See the graph. (ii) $y = 2x - 1$

(b) (i) See the graph. (ii) $y = {}^-\frac{1}{2}x + 4$

(c) $x = 2, y = 3$

The two lines intersect at this point.

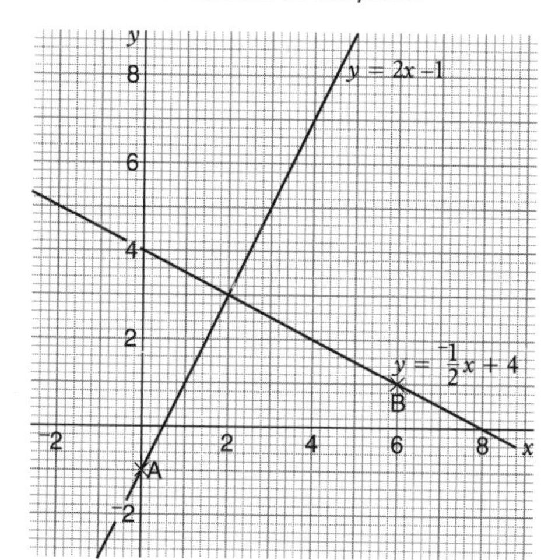

2 (a) graph (iv) (b) graph (ii) (c) graph (v)

3 (a) $\frac{1}{x} = x + 2$

(b) $x^2 + 2x - 1 = 0$

$$x = \frac{{}^-2 \pm \sqrt{(4+4)}}{2} = \frac{{}^-2 \pm \sqrt{8}}{2} = \frac{{}^-2 \pm 2\sqrt{2}}{2}$$

$= {}^-1 \pm \sqrt{2}$

$= {}^-2{\cdot}41$ or $0{\cdot}41$ (to 2 d.p.)

4 *In questions like these make both sides of the equation powers of the same number and then equate the powers.*

(a) $9^{\frac{1}{x}} = 3^{\frac{2}{x}} = 3^1$; solving $\frac{2}{x} = 1$ gives $x = 2$

(b) $16^x = 2^{4x} = 2^1$; solving $4x = 1$ gives $x = \frac{1}{4}$

(c) $81^{\frac{1}{4}} = 3^{4 \times \frac{1}{4}} = 3^1$; so $x = 1$

(d) $125^x = 5^{3x} = 5^{{}^-1}$; so $x = {}^-\frac{1}{3}$

5 (a) $n^2 - n = n(n-1)$

(b) Either n or $n - 1$ must be even, so the expression must always be divisible by 2 when n is a whole number.

6 $n^{\frac{4}{3}} = (\frac{1}{4})^4 = \frac{1}{256}$

$n^{\frac{1}{3}} = \frac{1}{4}$, so $n = (\frac{1}{4})^3 = \frac{1}{64}$ and $n^{{}^-1} = 64$

7 (a) Since the points $(7, 1)$ and $({}^-3, {}^-4)$ lie on the line we can write

$1 = 7m + c$ and ${}^-4 = {}^-3m + c$

(b) $m = \frac{1}{2}$, $c = {}^-2\frac{1}{2}$

(c) The equation of the line is $y = \frac{1}{2}x - 2\frac{1}{2}$.

Substituting $x = 12$ gives $y = 3\frac{1}{2}$.

8 (a) *In questions like this it is usually best to start by making T the subject and then substituting.*

$$\frac{1}{T} = \frac{S + R}{RS}$$

$$T = \frac{RS}{S + R} = \frac{32 \times {}^-50}{{}^-18} = 88{\cdot}\dot{8}$$

(b)
$$T = \frac{\frac{13}{4} \times \frac{13}{2}}{\frac{13}{4} + \frac{13}{2}} = \frac{\frac{13}{4} \times \frac{13}{2}}{\frac{13 + 26}{4}}$$

$$= \frac{\cancel{13}}{\cancel{4}} \times \frac{13}{2} \times \frac{\cancel{4}}{\cancel{39}_3} = \frac{13}{6} \text{ or } 2\frac{1}{6}$$

9 (a) See the curve.

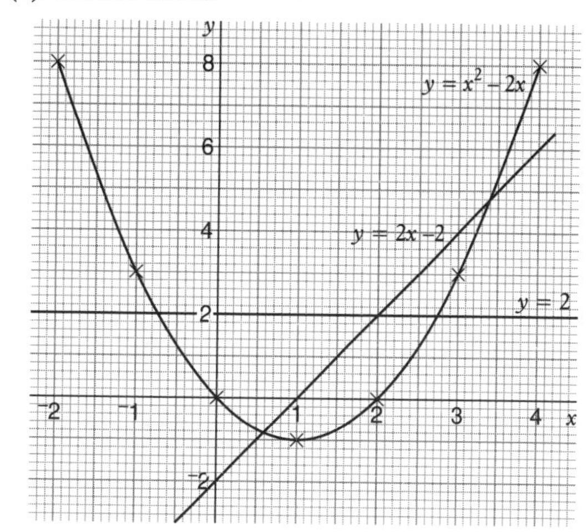

(b) *Draw the line $y = 2x - 2$ and find the x-coordinates of its intersection with the curve.*

$x = 3{\cdot}4$ or $0{\cdot}6$ (to 1 d.p.)

Answers from your graph are only likely to be accurate to 1 decimal place.

(c) *The line $y = 2$ cuts the curve at $x = {}^-0{\cdot}7$ and $2{\cdot}7$ (to 1 d.p.) Substitute values in $x^2 - 2x$ to solve the inequality.*

$x \leq {}^-0{\cdot}7$ or $x \geq 2{\cdot}7$

A more accurate solution, found algebraically, is $x \leq {}^-0{\cdot}73$ or $x \geq 2{\cdot}73$.

10 (a) (i) $\dfrac{300}{x}$ (ii) $\dfrac{210}{x+4}$

(b) $\dfrac{300(x+4)+210x}{x(x+4)} = \dfrac{510x+1200}{x(x+4)}$ seconds

(c) (i) $130x(x+4) = 510x + 1200$ which simplifies to
$130x^2 + 10x - 1200 = 0$ or
$13x^2 + x - 120 = 0$

(ii) The positive solution is $x = 3$ m/s.
Solve the quadratic by using the standard formula or by factorising, $(13x+40)(x-3)$.

11 *Draw up a table of values and then draw a smooth curve. Notice the symmetry of the values.*

t	0	1	2	3	4	0·5	3·5
h	0	15	20	15	0	8·75	8·75

(a) (i) Between 0·6 and 0·8 seconds
(ii) 20 metres

(b) *You have drawn a distance–time graph so the gradient of a tangent drawn at $h = 12$ will give an estimate for the speed.*
Between 10 m/s and 14 m/s.

(c) Rewrite $20t - 5t^2 = 12$ as $5t^2 - 20t + 12 = 0$ and use the quadratic formula to solve the equation.

$t = \dfrac{20 \pm \sqrt{(400-240)}}{10} = 2 \pm 0{\cdot}4\sqrt{10}$

$= 3{\cdot}26$ (to 2 d.p.)

Do not waste time working out both values of t. You need the larger value for the second time.

12 (a) (i) and (ii) An approximate solution to $x^2 + 2x - 4 = x^3$ is found by drawing the graph of $y = x^3$ and finding its points of intersection with $y = x^2 + 2x - 4$.
The curves cross once, so there is only one solution, $x = {}^-1{\cdot}7$ (to 1 d.p.).

(b) (i) The tangent at the point $(0, {}^-4)$ has gradient 2.
Look at the scales on the axes carefully.

(ii) A line parallel to the x-axis has zero gradient; the tangent at the lowest part of the curve (parabola) is parallel to the x-axis. This is the point where $x = {}^-1$.

(iii) $^-3$
One method is to put a ruler on the graph with a gradient of $^-1$ and then slide it, so it touches the curve.

13 (a) (i) The gradient of the line is $^-\frac{2}{3}$ and its y-intercept is 3, so the equation of the line is $y = {}^-\frac{2}{3}x + 3$.

(ii) Multiplying throughout by 3 to remove the fraction gives $3y = {}^-2x + 9$ which can be rearranged as $3y + 2x = 9$.

(b)

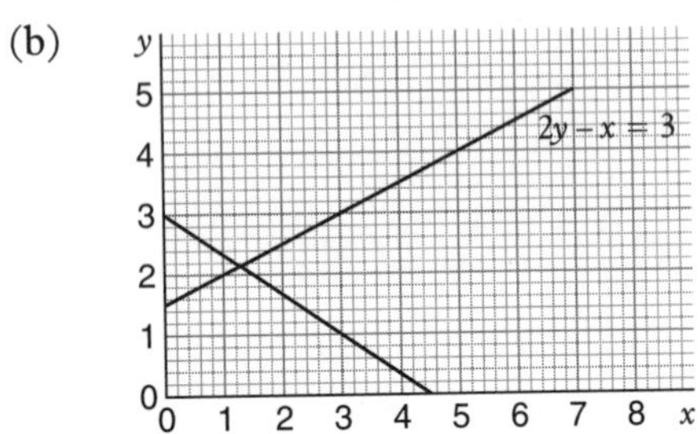

(c) The lines meet at the point (1·3, 2·1) (to 1 d.p.).
You should be able to take readings from your graph correct to 1 decimal place.

(d) $x = 1\frac{2}{7}$, $y = 2\frac{1}{7}$
Show all your working because you could gain method marks for it even if your answer was wrong. In this question you would lose a mark if you gave your final answer as a decimal.

14 *'Velocity' is another word for 'speed'.*

(a) Any answer between 11·3 m/s and 11·5 m/s is acceptable.

(b) The car accelerates steadily for about 10 seconds and then accelerates more gently (reduces its rate of acceleration) for the next 10 seconds. It then travels at a constant speed of 12 m/s for 30 seconds before decelerating steadily and coming to rest 15 seconds later.

(c) Between 0·6 m/s² and 0·7 m/s².
The gradient of the tangent at $t = 10$ gives the acceleration.

(d) The total distance travelled in the first 20 seconds is given by the area under the curve in this time interval.

Area
$\approx \frac{1}{2} \times 10 \times 9 + \frac{5}{2}(9 + 11{\cdot}5) + \frac{5}{2}(11{\cdot}5 + 12)$
$= 155$

So the distance $\approx$ 160 metres.

SHAPE, SPACE AND MEASURES

Angles (page 45)

Note: there are usually several ways to find an unknown angle in a diagram. So you could use different solutions to the ones here.
To help you, we have given reasons even where the question does not ask for them.

1 (a) (i) Each external angle of a regular pentagon is $360° \div 5 = 72°$.
So $\angle PQR = 180° - 72° = 108°$.

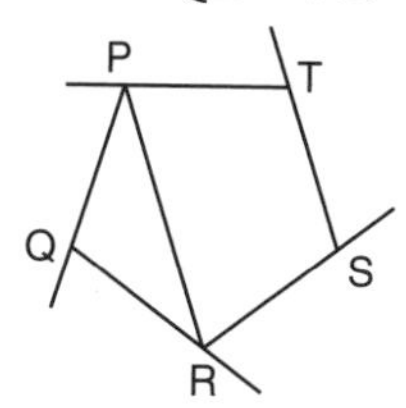

(ii) ΔPQR is isosceles, so $\angle PRQ = 72° \div 2 = 36°$.

(b) Triangles TQS and RSP are congruent and so are triangles PQR, STP, TPQ and QRS.

2 Where the interior angle is 155°, the exterior angle must be $180° - 155° = 25°$.
Where the interior angle is 165°, the exterior angle must be 15°.
So together, these two corners account for 40° of the total of the exterior angles, which sum to 360°.
So there are 320° left to account for.
Each of the other exterior angles is $180° - 170° = 10°$.
So there are another 32 ($320° \div 10°$) corners, that is a total of 34 corners in all.
So the polygon has 34 sides.

3 Triangles ABC and ADC are congruent (SAS).
So $\angle ADC = \angle ABC = 110°$ and
$\angle BCD = 360° - 110° - 110° - 60° = 80°$
(angle sum of quadrilateral).

4 (a) OR and OQ are radii, so ΔROQ is isosceles.
So $\angle RQO = 38°$.
So $\angle ROQ = 180° - 38° - 38° = 104°$.

(b) $\angle OQP = 90°$ (OQ is a radius, and QP is a tangent.)
So $\angle RQP = 90° - \angle OQR$
$= 90° - 38° = 52°$.

(c) PQ = PR (equal tangents)
So ΔPQR is isosceles.
So $\angle PRQ = \angle RQP = 52°$.
So $\angle RPQ = 180° - 52° - 52° = 76°$.

5 (a) $x = 160°$
(angle at centre = twice that at circumference)

(b) $\angle OAD = \angle OBD = 90°$
(each is an angle between a radius and the tangent)
So $y = 360° - 90° - 90° - 160° = 20°$
(angle sum of quadrilateral DAOB).

6 (a) $\angle BAC = \angle BDC = 43°$ (angles in same segment)

(b) $\angle DAB = 90°$ (angle in a semicircle)
So $\angle DAC = 90° - \angle BAC = 47°$.

(c) $\angle ACD = \angle ABD = 62°$ (angles in same segment)
$\angle AOD = 2 \times \angle ACD$
(angle at centre = twice angle at circumference)
So $\angle AOD = 124°$.

(d) ΔAOD is isosceles (OA and OD radii).
So $\angle OAD = \angle ODA$.
But $\angle ODA = 90° - 62° = 28°$ (angle sum of triangle ABD).
So $\angle OAD = 28°$.
But $\angle OAC + \angle OAD + \angle BAC = 90°$.
So $\angle OAC = 90° - 28° - 43° = 19°$.

7 (a) $\angle LPN = 80°$ (half angle at centre)

(b) Triangles MLO and MNO are congruent (SSS).
So $\angle LMO = \angle OMN = \frac{1}{2} \angle LMN = 40°$

(c) $\angle OLM = \angle LMO = 40°$
(base angles of isosceles triangle)

(d) $\angle MOL = 180° - 40° - 40° = 100°$
(angle sum of triangle is 180°)

8 $v = 90°$ (angle in a semicircle)
$w = 42°$ (base angle of isosceles ΔOFE)
$x = 84°$ ($\angle BOF$, angle at centre = twice $\angle BEF$, angle at circumference)
$y = 90°$ (angle between tangent and diameter)
$z = 48°$ (angle sum of ΔABE)

9 (a) (i) OA and OB are radii, so ΔOAB is isosceles.
So $\angle OBA = \angle OAB = p$.
So $\angle AOB = 180° - 2p$.
$\angle DOB = 180° - AOB$ (angles on a straight line)
$= 180° - (180° - 2p) = 2p$
Similarly $\angle DOC = 2q$.
So $\angle BOC = 2p + 2q$.

(ii) $\angle BAC = p + q$, so $\angle BOC = 2\angle BAC$.

(b) $\angle DOB = 2p$ (proof exactly as before) and $\angle DOC = 2q$.
$\angle BOC = 2q - 2p$ and $\angle BAC = q - p$.
So we see that $\angle BOC = 2\angle BAC$.

10 (a) $\angle ADC = 90°$, since ABCD is a rectangle.
The line AC subtends $\angle ADC$ at the circumference of the circle, that is it subtends 90°.
Therefore ΔADC must lie in a semicircle, so AC is a diameter.

(b) (i) TC = TD (equal tangents)
So ΔTDC is isosceles.
Therefore $\angle TCD = \angle TDC = 42°$.
$\angle CTD = 180° - 42° - 42°$ (angle sum of triangle)
so $\angle CTD = 96°$.

(ii) $\angle ACT = 90°$ (angle between tangent and diameter AC)
Therefore $\angle ACD = 90° - 42° = 48°$.
AB is parallel to DC (opposite sides of rectangle).
So $\angle CAB = \angle ACD = 48°$ (alternate angles).

11 (a) (i) $\angle COB = \angle OBD = 50°$
(alternate angles, OC parallel to DB).

(ii) OB = OC (radii), so ΔOCB is isosceles.
So $\angle OCB = \angle OBC$.
So $50° + 2\angle OBC = 180°$ (angle sum of ΔOCB)
so $2\angle OBC = 130°$ and $\angle OBC = 65°$.

(iii) OB is perpendicular to BX (tangent and radius), so
$\angle CBX = 90° - \angle CBO = 25°$.

(iv) $\angle ACB = 90°$ (lies in a semicircle)
So $\angle BCX = 90°$.
So $\angle BXC = 180° - 90° - 25° = 65°$.

(v) $\angle CAB = \frac{1}{2}\angle COB$
(angle at centre twice that at circumference)
So $\angle CAB = 25°$.

(b) Suppose $\angle CAB = a$.
Now OA = OC (radii),
so ΔOAC is isosceles.
So $\angle OAC = \angle OCA = a$.
Also, $\angle BDC = \angle CAB = a$
(angles in the same segment).
OC is parallel to DB,
so $\angle OCD = \angle CDB$ (alternate angles),
and $\angle OCD = a$.

Thus OC bisects $\angle ACD$.

Note that you could prove that both $\angle OCA$ and $\angle OCD$ are 25° in this example; but the proof will hold whatever the size of the given angle in the diagram.

12 (a) Each external angle is $360° \div 5 = 72°$.
So each internal angle is $180° - 72° = 108°$.

(b) 90°

(c) Where the pentagons and squares meet, the interior angles must add up to 360° (angles at a point).
It is obvious that no combination of 108° and 90° can add up to 360° (except 90° + 90° + 90° + 90°, which is the equivalent of four squares tessellating without any pentagons!).

More help or practice

Review of angle relationships ► Book Y4 pages 25 to 27
Calculating angles of regular polygons ► Book Y4 pages 29 to 32
Angle properties of a circle ► Book YX1 pages 1 to 11
Deduction (proof) ► Book YX1 pages 12 to 16
Technical terms of a circle ► Book YX1 pages 81 to 82
Properties of tangents to a circle ► Book YX1 pages 83 to 87
Conditions for congruent triangles ► Book YR+ pages 14 to 18, Book YX1 pages 63 to 69

Pythagoras and trigonometry in two dimensions (page 49)

*In questions like these you might need to give a rounded answer to one part of a question and then use the answer in another part. If so, remember to use the **unrounded** answer in your calculation.*

1 By Pythagoras, $HC^2 + 350^2 = 521^2$.
$HC = \sqrt{(521^2 - 350^2)}\,m = 385{\cdot}928\ldots m$
$= 386\,m$ (to nearest metre)
So Mohamed saves $(386 + 350 - 521)\,m = 215\,m$ by taking the short cut.

2

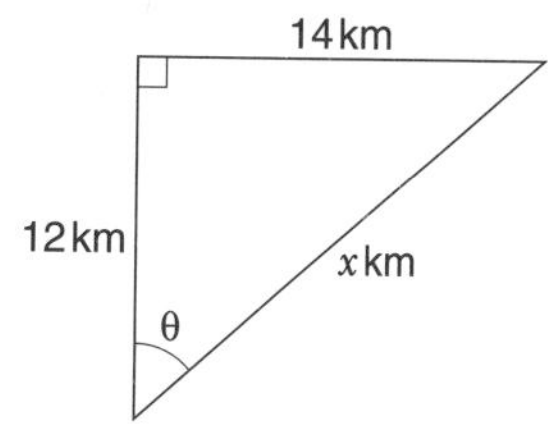

(a) $x^2 = \sqrt{(12^2 + 14^2)} = 18{\cdot}439\ldots$
Distance = 18 km (to nearest km)

(b) $\tan\theta = \frac{14}{12}$, $\theta = 49{\cdot}398\ldots°$
Bearing = 049° (to nearest degree)
Bearings are always measured clockwise from north.

3 (a) $\frac{h}{6} = \sin 78°$,
$h = 6\sin 78°$
$h = 5{\cdot}8688\ldots$
Height = 5·9 m (to 2 s.f.)

(b) $\cos\theta = 1{\cdot}6 \div 6$,
$\theta = 74{\cdot}533\ldots°$
$\theta = 75°$ (to 2 s.f.)

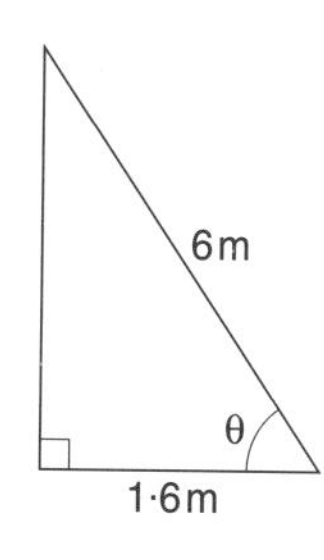

4

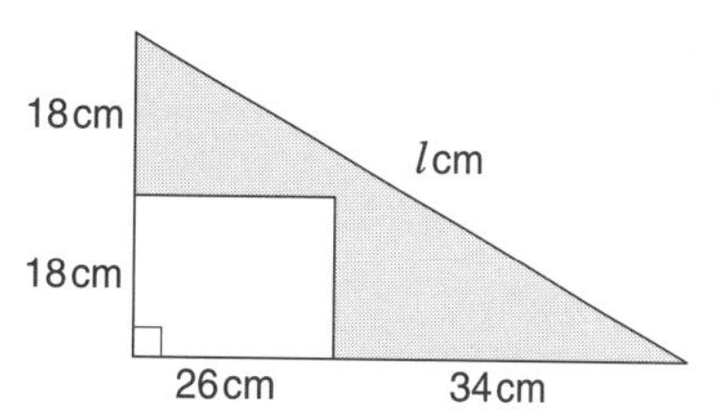

(a) The known sides of the triangle above are 36 cm and 60 cm.
So $l^2 = 36^2 + 60^2$ and
$l = \sqrt{(36^2 + 60^2)} = 69{\cdot}971\ldots$
Length = 70 cm (to nearest cm)

(b) Shaded area $= \frac{1}{2}((36 \times 60) - (26 \times 18))\,cm^2$
$= (1080 - 468)\,cm^2 = 612\,cm^2$
$= 610\,cm^2$ (to 2 s.f.)

(c) $\tan x = \frac{36}{60}$
$x = 30{\cdot}963\ldots° = 31°$ (to 2 s.f.)

(d)

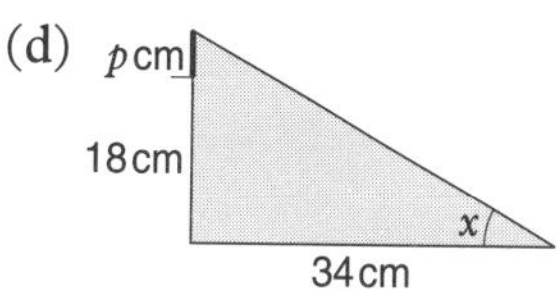

$(p + 18) = 34\tan x = 34 \times \frac{36}{60}$ (from above)
$p = (20{\cdot}4 - 18)\,cm = 2{\cdot}4\,cm$
You could use similar triangles to find p, since the triangle shown immediately above is similar to the large triangle in part (a).

5

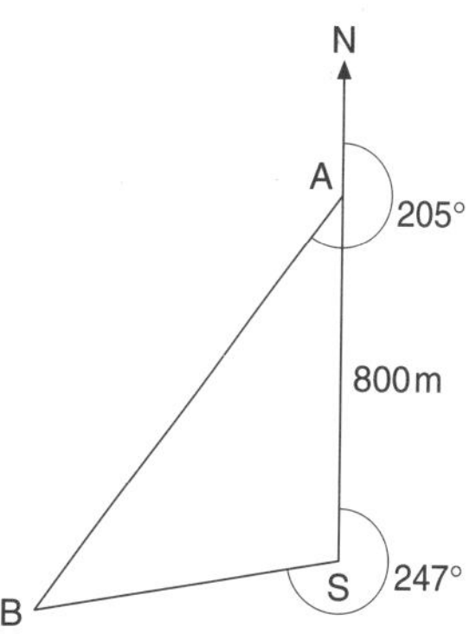

(a) In ΔABS, $\angle S = 360° - 247° = 113°$.
Also $\angle BAS = 205° - 180° = 25°$.
So $\angle ABS = 180° - 113° - 25° = 42°$.

(b) $\frac{AB}{\sin S} = \frac{800}{\sin B}$ *using the sine rule*

$AB = \sin 113° \times \frac{800}{\sin 42°}$
$= 1100{\cdot}53\ldots m$
$= 1100\,m$ (to 2 s.f.)

6

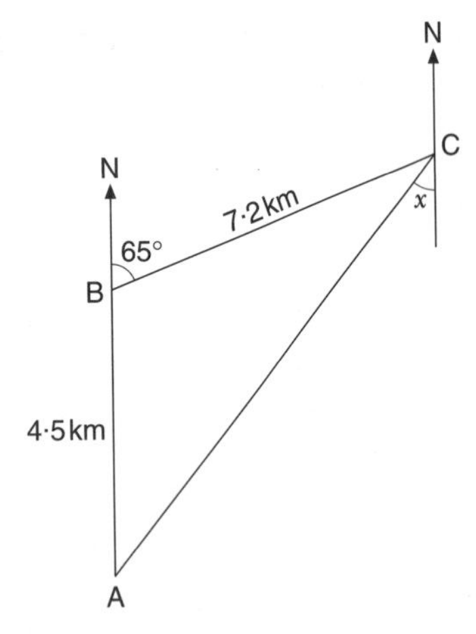

(a) $\angle ABC = 180° - 65° = 115°$

$AC^2 = 4{\cdot}5^2 + 7{\cdot}2^2 - 2 \times 4{\cdot}5 \times 7{\cdot}2 \times \cos 115°$

$= 20{\cdot}25 + 51{\cdot}84 + 27{\cdot}385\ldots$

$= 99{\cdot}475\ldots$

$AC = 9{\cdot}973\ldots \text{km} = 10{\cdot}0\,\text{km}$ (to 1 d.p.)

(b) $\dfrac{7{\cdot}2}{\sin A} = \dfrac{9{\cdot}973\ldots}{\sin 115°}$

$\sin A = \dfrac{7{\cdot}2 \times \sin 115°}{9{\cdot}973\ldots} = 0{\cdot}654\ldots$

$A = 40{\cdot}863\ldots°$

Bearing of A from C $= 180° + x$.

But $x = A$ *alternate angles, AB parallel to CD*

So bearing $= 180° + 40{\cdot}863\ldots°$

$= 220{\cdot}863\ldots° = 221°$ (to nearest degree)

7 (a) $R = 180° - 48° - 57° = 75°$.

$\dfrac{QR}{\sin P} = \dfrac{120}{\sin R}$

$QR = \sin 57° \times \dfrac{120}{\sin 75°}$

$= 104{\cdot}190\ldots \text{m}$

$\dfrac{PR}{\sin Q} = \dfrac{120}{\sin R}$

$PR = \sin 48° \times \dfrac{120}{\sin 75°} = 92{\cdot}323\ldots \text{m}$

So Jeff needs $(104{\cdot}190\ldots + 92{\cdot}323\ldots)$m

$= 196{\cdot}513\ldots$m

$= 197$m (to nearest metre above)

(b) Area $= \frac{1}{2} PR \times PQ \times \sin 57°\,\text{m}^2$

$= \frac{1}{2} \times 92{\cdot}323\ldots \times 120 \times \sin 57°\,\text{m}^2$

$= 4645{\cdot}725\ldots \text{m}^2 = 4650\,\text{m}^2$ (to 3 s.f.)

8 (a)

In ΔOXY, $\angle YOX = 310° - 274° = 36°$.

So area of ΔOXY

$= (\frac{1}{2} \times OX \times OY \times \sin 36°)\,\text{km}^2$

$= (\frac{1}{2} \times 85 \times 65 \times \sin 36°)\,\text{km}^2$

$= 1623{\cdot}7\ldots \text{km}^2 = 1600\,\text{km}^2$ (to 2 s.f.)

(b)

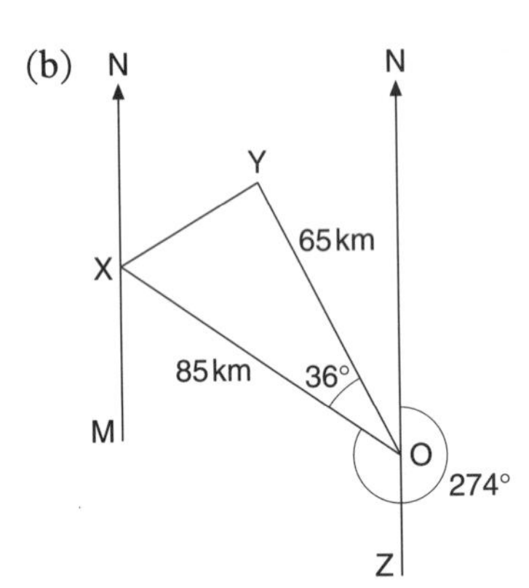

We need to find ∠OXY.

To find this angle, we first find XY.

$XY^2 = XO^2 + YO^2 - 2 \times XO \times YO \times \cos 36°$

$= 85^2 + 65^2 - 2 \times 85 \times 65 \times \cos 36°$

$= 2510{\cdot}362\ldots$

$XY = 50{\cdot}103\ldots \text{km}$

Now $\dfrac{65}{\sin OXY} = \dfrac{XY}{\sin XOY}$

$\sin OXY = \dfrac{65 \sin 36°}{50{\cdot}103\ldots}$ and $\angle OXY = 49{\cdot}68\ldots°$

Now $\angle ZOX = 274° - 180° = 94°$

So $\angle MXO = 180° - 94°$ *alternate angles*

$= 86°$.

So $\angle NXY = 180° - 49{\cdot}68\ldots° - 86° = 44{\cdot}31\ldots°$

Bearing of Y from X = 044°

(to nearest degree)

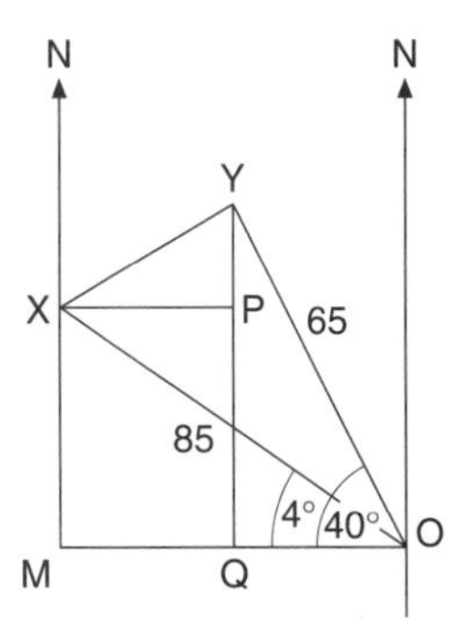

An alternative solution to part (b) is to find angle YXP above.

$YP = YQ - XM = 65\sin 40° - 85\sin 4°$
$= 35{\cdot}851...$

$XP = MO - QO = 85\cos 4° - 65\cos 40°$
$= 35{\cdot}000...$

So $\tan YXP = \dfrac{YP}{XP} = \dfrac{35{\cdot}851...}{35{\cdot}000...}$

$YXP = 45{\cdot}68...°$

Bearing $= 90° - 45{\cdot}68...° = 44{\cdot}31...°$
$= 044°$ (to nearest degree)

More help or practice

Pythagoras' rule ► Book Y3 pages 102 to 107
Tangents and their inverses ► Book Y2 pages 26 to 33
Sines and cosines and their inverses
► Book Y2 pages 55 to 64, Book Y3 pages 112 to 114
The sine rule ► Book YX2 pages 19 to 26
The cosine rule ► Book YX2 pages 58 to 64
Area of a triangle ► Book YX2 page 140

Pythagoras and trigonometry in three dimensions (page 52)

1 (a) In ΔABC, $AC^2 = AB^2 + BC^2$
$AC^2 = 60^2 + 25^2 = 4225$
$AC = \sqrt{4225}\,\text{cm} = 65\,\text{cm}$

(b) In ΔACG, $AG^2 = AC^2 + CG^2$
$CG^2 = 72^2 - 65^2 = 959$
$CG = \sqrt{959} = 30{\cdot}96...\,\text{cm}$
$= 31\,\text{cm}$ (to 2 s.f.)

(c) Angle required = HAG

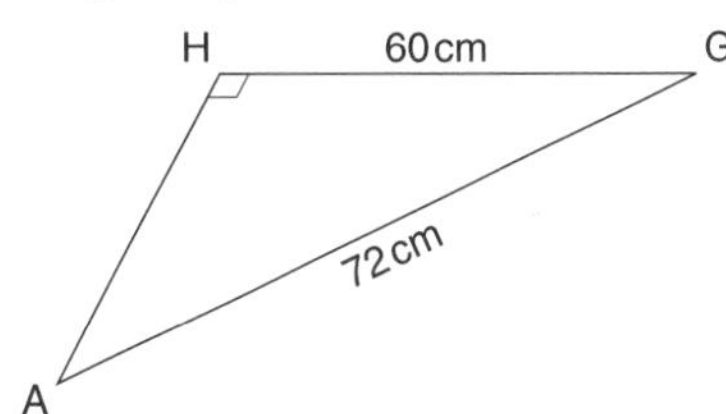

$\sin HAG = \frac{60}{72}$
$\angle HAG = 56{\cdot}44...°$
$= 56°$ (to nearest degree)

2 (a) In ΔABC, $AC = 47\sin 35°\,\text{m}$
$= 26{\cdot}95...\,\text{m} = 27\,\text{m}$ (to 2 s.f.)

(b) In ΔACT, $CT = AC\tan 18°$
$= 8{\cdot}759...°\,\text{m} = 8{\cdot}8\,\text{m}$ (to 2 s.f.)

3 (a) In ΔMPT, $MT = 70 = PT\tan 32°$

So $PT = \dfrac{70}{\tan 32°} = 112{\cdot}02...\,\text{m}$.

In ΔMTQ, $QT = \dfrac{70}{\tan 35°} = 99{\cdot}97...\,\text{m}$.

In ΔTPQ, using the cosine rule,
$PQ^2 = PT^2 + QT^2 - 2 \times PT \times QT \times \cos 134°$
$= 38\,102{\cdot}3...$
$PQ = 195{\cdot}19...\,\text{m} \approx 195\,\text{m}$

(b) In ΔTPQ, using the sine rule,

$\dfrac{PQ}{\sin PTQ} = \dfrac{QT}{\sin TPQ}$

$\dfrac{195{\cdot}19...}{\sin 134°} = \dfrac{99{\cdot}97...}{\sin TPQ}$

$\sin TPQ = \dfrac{99{\cdot}97...\sin 134°}{195{\cdot}19...} = 0{\cdot}368...$

$\angle TPQ = 21{\cdot}61...° = 22°$ (to nearest degree)

4 The sketch below shows a cross-section through the middle line of the roof:

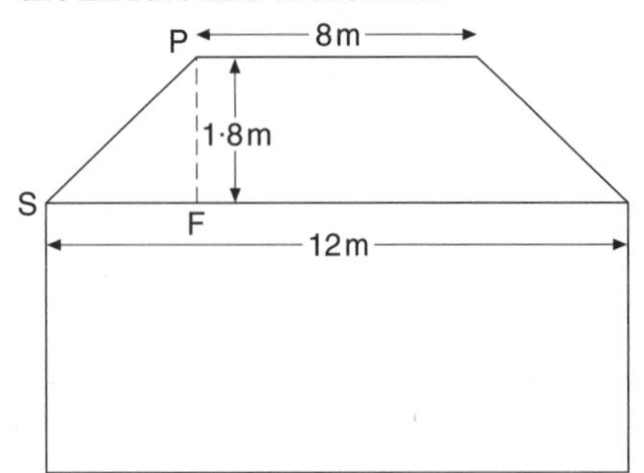

F is the foot of the perpendicular from P onto the base plane of the roof, and ∠PSF is the angle that PS makes with the horizontal.
Since the house is symmetrical, we have
$SF = \frac{1}{2}(12 - 8)\text{m} = 2\text{m}$.

In ΔPFS, $\tan PSF = \frac{1{\cdot}8}{2} = 0{\cdot}9$ and

$\angle PSF = 41{\cdot}987\ldots^\circ = 42^\circ$ (to nearest degree)

5 (a)

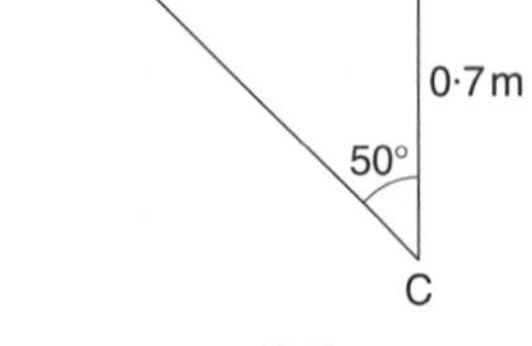

In ΔCDB, $\frac{0{\cdot}7}{CD} = \cos 50^\circ$

$CD = \frac{0{\cdot}7}{\cos 50^\circ} = 1{\cdot}089\ldots\text{m}$

Length of arms = 1·1 m (to 2 s.f.)

(b) (i) In ΔCDB, $DB = 0{\cdot}7 \tan 50^\circ = 0{\cdot}8342\ldots\text{m}$
$= 0{\cdot}83\text{m}$ (to 2 s.f.)

Alternatively you could have used Pythagoras' rule.

(ii) Looking at the top view of the whole line, we see:

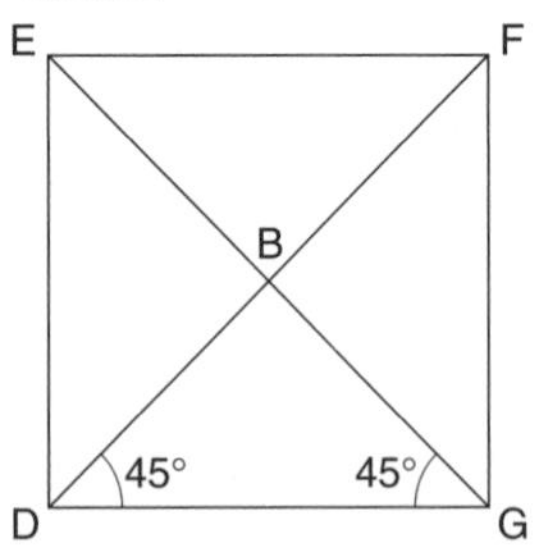

Since DEFG is a square, the angles in DBG will be 45°, with an angle of 90° at B.
We know that DB = 0·8342... m, and that DB = BG.
So $DG^2 = DB^2 + BG^2 = 2DB^2$
$= 2 \times (0{\cdot}8342\ldots)^2 = 1{\cdot}3918\ldots$
$DG = 1{\cdot}1797\ldots\text{m}$
Total length of clothes line $= 4 \times DG$
$= 4{\cdot}719\ldots\text{m} = 4{\cdot}72\text{m}$ (to 3 s.f.)

More help or practice

Using Pythagoras' rule in three dimensions
► Book Y3 pages 109 to 111
Angles in three dimensions ► Book Y4 pages 134 to 140

Length, area and volume 1 (page 54)

1 (a) Length = 2 straight pieces plus semicircle.
Straight pieces length $= 2 \times 65\,\text{cm} = 130\,\text{cm}$
Semicircle $= \frac{1}{2} \times 2\pi \times 30\,\text{cm} = 94{\cdot}24\ldots\text{cm}$
Length of hoop $= 130 + 94{\cdot}24\ldots\text{cm}$
$= 224{\cdot}24\ldots\text{cm} = 220\,\text{cm}$ (to 2 s.f.)

(b) *It is probably best to work in metres in this part, as an answer in cm² will be large.*
Area required
= 2 flat strips plus half a cylinder
Total area of flat strips
$= 2 \times 0{\cdot}45 \times 1{\cdot}5\,\text{m}^2 = 1{\cdot}35\,\text{m}^2$
Area of half cylinder
$= \frac{1}{2} \times 2 \times \pi \times 0{\cdot}30 \times 1{\cdot}5\,\text{m}^2$
$= 1{\cdot}413\ldots\text{m}^2$
Total area
$= (1{\cdot}35 + 1{\cdot}413\ldots)\,\text{m}^2$
$= 2{\cdot}763\ldots\text{m}^2$
$= 2{\cdot}8\,\text{m}^2$ (to 2 s.f.) or $2{\cdot}76\,\text{m}^2$ (to 3 s.f.)
Alternatively, you could say that the required area is equal to the length of one hoop above ground, multiplied by the length of the cloche.
From (a), length of hoop $= (224{\cdot}24\ldots - 40)\,\text{cm}$
$= 184{\cdot}24\ldots\text{cm} = 1{\cdot}8424\ldots\text{m}$.
Area $= 1{\cdot}5 \times 1{\cdot}8424\ldots\text{m}^2$
$= 2{\cdot}763\ldots\text{m}^2$
$= 2{\cdot}8\,\text{m}^2$ (to 2 s.f.) or $2{\cdot}76\,\text{m}^2$ (to 3 s.f.)

2 (a) *Mug A*
If the height is h cm, then the volume is
$7{\cdot}5 \times 7{\cdot}5 \times h\,\text{cm}^3 = 56{\cdot}25h\,\text{cm}^3$.
But we know the volume = 0·45 litres
$= 450\,\text{cm}^3$.
So $h = 450 \div 56{\cdot}25 = 8$.
Height = 8 cm

Mug B
The base is a regular hexagon. Think of this as 6 equilateral triangles.
Each triangle has sides 4 cm long, and angles of 60°.
So the area of each triangle $= \frac{1}{2}ab \sin C$
$= \frac{1}{2} \times 4 \times 4 \times \sin 60° = 6{\cdot}928\ldots\text{cm}^2$
Area of base $= 6 \times 6{\cdot}928\ldots\text{cm}^2$
$= 41{\cdot}569\ldots\text{cm}^2$
So the height $= 450 \div 41{\cdot}569\ldots\text{cm}$
$= 10{\cdot}825\ldots\text{cm}$
$= 11\,\text{cm}$ (to 2 s.f.)

(b) Let the radius of the mug be r cm.
The height of the mug is 10 cm and the volume is 450 cm³.
So $\pi r^2 \times 10 = 450$ and
$r = \sqrt{\dfrac{450}{10\pi}} = 3{\cdot}784\ldots\text{cm} = 3{\cdot}8\,\text{cm}$ (to 2 s.f.)

3 *The end of the case can be thought of as four right-angled triangles.*

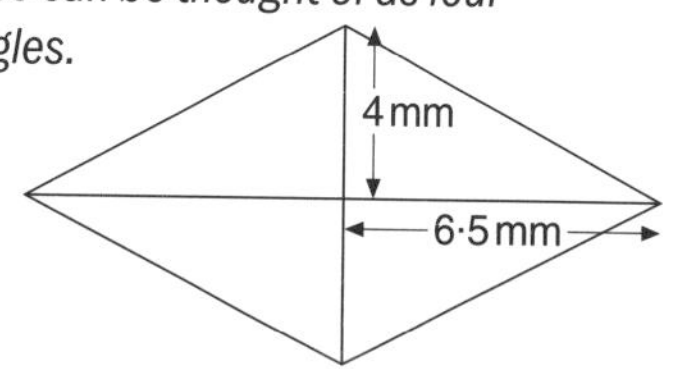

Area of each triangle
$= \frac{1}{2} \times 4 \times 6{\cdot}5\,\text{mm}^2 = 13\,\text{mm}^2$,
so area of end $= 4 \times 13\,\text{mm}^2 = 52\,\text{mm}^2$.
The case is a prism, so its volume
= area of end × length $= 71 \times 52\,\text{mm}^3$
$= 3692\,\text{mm}^3 = 3700\,\text{mm}^3$ (to 2 s.f.)
You could work in cm throughout this question, when the answer will be 3·7 cm³. If you do this, be sure to show your units clearly in your answer.

4 (a) The circle has diameter 7·1 cm, so the radius is 3·55 cm.
Area $= \pi \times 3{\cdot}55^2\,\text{cm}^2 = 39{\cdot}59\ldots\text{cm}^2$
$= 40\,\text{cm}^2$ (to 2 s.f.) or $39{\cdot}6\,\text{cm}^2$ (to 3 s.f.)

(b) Area of trapezium
$= \frac{1}{2}(5 + 10) \times 7{\cdot}1\,\text{cm}^2 = 53{\cdot}25\,\text{cm}^2$
Shaded area
= area of trapezium − area of circle
$= (53{\cdot}25 - 39{\cdot}59\ldots)\,\text{cm}^2 = 13{\cdot}66\ldots\text{cm}^2$
$= 14\,\text{cm}^2$ (to 2 s.f.) or $13{\cdot}7\,\text{cm}^2$ (to 3 s.f.)
Make sure you use the ***unrounded*** *area of the circle you found in (a) when working part (b).*

5 The area of the wok will have dimensions of (length)². $\frac{4}{7}\pi r^2$ is the only expression with these dimensions, so this is the only one that could be the area.

> **More help or practice**
> Area of a parallelogram, triangle and trapezium ► Book Y2 pages 113 to 118
> Volume of a cuboid and prism ► Book Y3 pages 119 to 121
> Volume of a cylinder ► Book Y3 page 122
> Distinguish between the formulas for perimeter, area and volume by considering dimensions ► Book Y5 pages 98 to 107

Length, area and volume 2 (page 56)

1 (a) Volume $= \frac{1}{3}\pi \times 8{\cdot}2^2 \times 14{\cdot}3\,\text{cm}^3$
$= 1006{\cdot}91\ldots\text{cm}^3 = 1000\,\text{cm}^3$ (to 2 s.f.)

(b) Slant height of cone
$= \sqrt{(8{\cdot}2^2 + 14{\cdot}3^2)}\,\text{cm} = 16{\cdot}48\ldots\text{cm}$
So curved surface area
$= \pi \times 8{\cdot}2 \times 16{\cdot}48\ldots\text{cm}^2$
$= 420\,\text{cm}^2$ (to 2 s.f.)

2 Volume required = volume of whole sphere – volume of space inside.
Radius of whole sphere = 40 mm,
so radius of space = (40 – 5) mm = 35 mm.
So required volume $= (\frac{4}{3}\pi \times 40^3 - \frac{4}{3}\pi \times 35^3)\,\text{mm}^3$
$= \frac{4}{3}\pi \times (40^3 - 35^3)\,\text{mm}^3$
$= 88\,488{\cdot}1\ldots\text{mm}^3$
$= 88\,\text{cm}^3$ (to 2 s.f.)

You could give the answer in mm³, but cm³ is probably a more sensible unit.
Remember to use the radius, not the diameter in the formula.

3 Volume of hemisphere $= \frac{1}{2} \times \frac{4}{3}\pi \times 3{\cdot}5^3\,\text{cm}^3$
$= 89{\cdot}79\ldots\text{cm}^3$
Volume of cone $= \frac{1}{3}\pi \times 3{\cdot}5^2 \times 11\,\text{cm}^3$
$= 141{\cdot}10\ldots\text{cm}^3$
So total volume $= 230{\cdot}9\ldots\text{cm}^3$
$= 230\,\text{cm}^3$ (to 2 s.f.)

Since the original data is given to 2 s.f. – we are told the radius is 3·5 cm, not 3·50 cm – you should certainly not give your answer to any more accuracy than 2 s.f. An answer to 1 s.f. could also be justified, on both practical and mathematical grounds.

4 (a) (i) Volume of hemispherical base
$= \frac{1}{2} \times \frac{4}{3}\pi \times 1{\cdot}3^3\,\text{cm}^3 = 4{\cdot}601\ldots\text{cm}^3$.
There is a cylinder of water above the hemisphere, of height $(11{\cdot}3 - 1{\cdot}3)$ cm = 10 cm, and radius 1·3 cm.
Volume of this cylinder
$= \pi \times 1{\cdot}3^2 \times 10\,\text{cm}^3$
$= 53{\cdot}09\ldots\text{cm}^3$
So total volume of water
$= 4{\cdot}601\ldots + 53{\cdot}09\ldots\text{cm}^3$
$= 57{\cdot}69\ldots\text{cm}^3 = 58\,\text{cm}^3$ (to 2 s.f.)

(ii) Area in contact with water
= curved surface area of hemisphere + curved surface area of cylinder
$= (\frac{1}{2} \times 4\pi \times 1{\cdot}3^2) + (2\pi \times 1{\cdot}3 \times 10)\,\text{cm}^2$
$= (10{\cdot}61\ldots + 81{\cdot}68\ldots)\,\text{cm}^2$
$= 92{\cdot}29\ldots\text{cm}^2$
$= 92\,\text{cm}^2$ (to 2 s.f.)

(b) Volume of stone = volume of water in a 24 mm high cylinder
24 mm = 2·4 cm,
so volume of cylinder $= \pi \times 1{\cdot}3^2 \times 2{\cdot}4\,\text{cm}^3$
$= 12{\cdot}74\ldots\text{cm}^3$
$= 13\,\text{cm}^3$ (to 2 s.f.)

5 (a) *The portion can be considered as a prism, whose cross-section is a sector.*
Area of sector $= \frac{40}{360}$ of the whole circle of which it is part
$= \frac{40}{360} \times \pi \times 5^2\,\text{cm}^2 = 8{\cdot}726\ldots\text{cm}^2$
Volume of portion $= \text{area} \times 1{\cdot}5\,\text{cm}^3$
$= 13{\cdot}08\ldots\text{cm}^3 = 13\,\text{cm}^3$ (to 2 s.f.)

(b) The area not covered by the label is a segment.
Its area = area of sector – area of label
$= (\frac{40}{360} \times \pi \times 5^2 - \frac{1}{2} \times 5 \times 5 \times \sin 40°)\,\text{cm}^2$
$= (8{\cdot}726\ldots - 8{\cdot}034\ldots)\,\text{cm}^2$
$= 0{\cdot}691\ldots\text{cm}^2 = 0{\cdot}69\,\text{cm}^2$ (to 2 s.f.)

6 (a) Area of sectors are in total

$\frac{40}{360} + \frac{60}{360} + \frac{40}{360} = \frac{140}{360}$ of the whole circle radius 24 cm.

So total area of sectors

$= \frac{140}{360} \times \pi \times 24^2\,\text{cm}^2 = 703{\cdot}7\ldots\text{cm}^2$

$= 700\,\text{cm}^2$ (to 2 s.f.)

(b) Total length of curved framing in sectors

$= \frac{140}{360} \times 2\pi \times 24\,\text{cm} = 58{\cdot}64\ldots\text{cm}$

So total length of framing

$= (6 \times 24 + 58{\cdot}64\ldots)\,\text{cm}$

$= 202{\cdot}6\ldots\text{cm}$

$= 2{\cdot}0\,\text{m}$ (to 1 d.p.)

7 (a) Triangles BAD and CED are similar,

so $\frac{AD}{AB} = \frac{ED}{EC}$.

$\frac{15}{3{\cdot}5} = \frac{ED}{2{\cdot}1}$, so $ED = \frac{2{\cdot}1 \times 15}{3{\cdot}5}\,\text{cm} = 9\,\text{cm}$

(b) Volume of cup

= volume of large cone
 – volume of small cone

$= (\frac{1}{3}\pi \times 3{\cdot}5^2 \times 15 - \frac{1}{3}\pi \times 2{\cdot}1^2 \times 9)\,\text{cm}^3$

$= 150{\cdot}8\ldots\,\text{cm}^3$

$= 150\,\text{cm}^3$ (to 2 s.f.)

Note: you could be more sophisticated, and say that since the radii of the cones are in the ratio 3·5:2·1 (= 5:3), the volumes are in the ratio 125:27. So the truncated cone has a volume that is $\frac{125-27}{125}$ of the large cone, and the cup has a volume of $\frac{98}{125} \times \frac{1}{3}\pi \times 3{\cdot}5^2 \times 15$

$= 150{\cdot}8\ldots\,\text{cm}^3$

$= 150\,\text{cm}^3$ (to 2 s.f.).

More help or practice

Volume of a pyramid and a cone ► Book Y3 page 123, Book Y5 pages 6 to 8

Volume and surface area of a sphere and a cone ► Book YX2 pages 38 to 39

Transformations (page 58)

1 (a) Reflection in the line CH

(b) Rotation through 90° ($\frac{1}{4}$ turn anti-clockwise) about D

2 There are many solutions. Here is one.

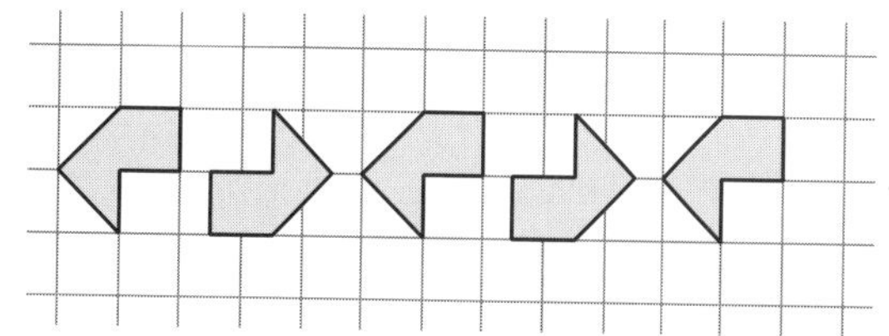

3 (a) Enlargment centre O, scale factor $^{-}\frac{5}{3}$

(b) $OB = \frac{3}{5} \times 7\,\text{cm} = 4{\cdot}2\,\text{cm}$

4 (a) $\begin{bmatrix} 3 \\ ^{-}3 \end{bmatrix}$

(b) $(0, ^{-}1)$

5 (a) and (b)

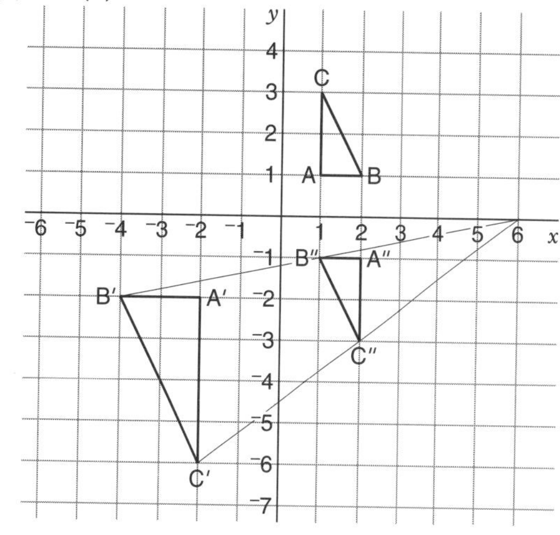

(c) Enlargement, scale factor $\frac{1}{2}$, centre (6, 0)

6 (a) Rotation through $^-90°$ ($\frac{1}{4}$ turn clockwise) centre O

(b) $(0, ^-3)$

(c) Translation 3 across and 3 down or $\begin{bmatrix} 3 \\ -3 \end{bmatrix}$

(d)

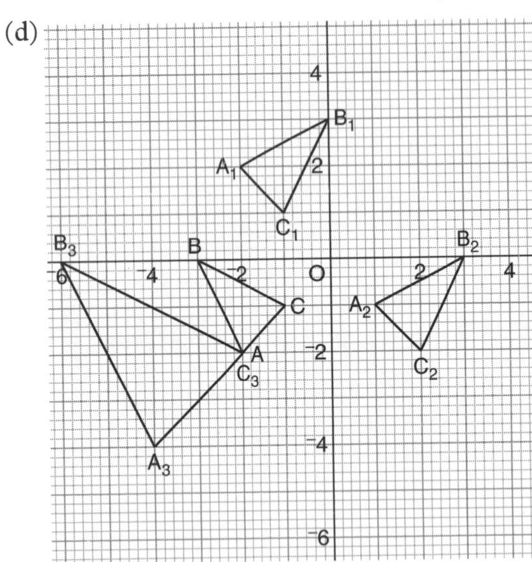

More help or practice

Column vectors and translations ► Book Y3 pages 27 to 31, 50

Rotations of 180° and 90° ► Book Y3 pages 46 to 49

Reflection ► Book Y3 pages 44 to 45

Recognising mappings ► Book Y3 pages 51 to 55

Enlargements and reductions
► Book Y1 pages 76 to 85, Book Y3 pages 8 to 10

1- and 2-way stretches ► Book Y3 pages 1 to 7

Strip patterns ► Book Y3 pages 32 to 33

Scales and similarity (page 60)

1 (a) 1 cm on the map is really 50 000 cm
= 500 m = 0·5 km on the ground.
So 12 cm is really 6 km.

(b) 1 cm stands for 0·5 km, so
$1\,\text{cm}^2$ stands for $0{\cdot}5 \times 0{\cdot}5\,\text{km}^2 = 0{\cdot}25\,\text{km}^2$.
So $20\,\text{cm}^2$ stands for $20 \times 0{\cdot}25\,\text{km}^2 = 5\,\text{km}^2$.

2 1 mm on the map = 8 000 000 mm on the ground
= 800 000 cm = 8000 m = 8 km.
So 100 km on the ground = 100 ÷ 8 mm
= 12·5 mm on the map.
Alternatively convert 100 km to mm and then divide by 8 000 000.

3 *First you need a sketch, supposing that Nefyn (N) and the tops of Snowdon (S) and Garnedd-goch (G) are in a straight line.*

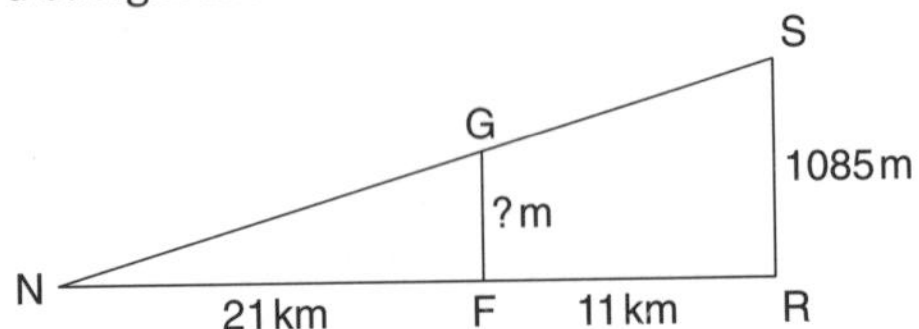

SRN is an enlargement of GFN, with scale factor
$\frac{32}{21} = 1{\cdot}523\ldots$
So GF = SR ÷ 1·523… = 712·03… m.
So Garnedd-goch must be at least 720 m high.
Note that we have rounded up here (because Garnedd-goch must be higher than GF), but it does not really matter much! An answer of 710 m would be quite acceptable.

4 (a) 1 cm on the map is 50 000 cm on the ground
= 500 m = 0·5 km.
So 2·3 cm on the map = 2·3 × 0·5 km on the ground = 1·15 km
= 1·2 km (to 2 s.f.).

(b) 1 cm represents 50 000 cm = 500 m, so
$1\,\text{cm}^2$ represents $500 \times 500\,\text{m}^2 = 250\,000\,\text{m}^2$
= 250 000 ÷ 10 000 hectares = 25 hectares.
So $2\,\text{cm}^2$ represents 2 × 25 hectares
= 50 hectares.

5 (a) The length scale factor from the small to the large model is $\frac{24}{16} = 1{\cdot}5$.
So the area scale factor is $1{\cdot}5^2$.
So the armour area $= 9 \times 1{\cdot}5^2 \text{cm}^2$
$= 20{\cdot}25 \text{cm}^2$
$= 20 \text{cm}^2$ (to 2 s.f.)

(b) Length scale factor $= 1{\cdot}5$, so volume scale factor $= 1{\cdot}5^3 = 3{\cdot}375$.
So weight (which is proportional to volume)
$= 270 \times 3{\cdot}375$ grams
$= 911{\cdot}25$ grams
$= 910$ grams (to 2 s.f.).

6 If the coins are similar, then the length scale factor $= \frac{28}{22} = 1{\cdot}2727\ldots$
So the volume factor will be
$(1{\cdot}2727\ldots)^3 = 2{\cdot}061\ldots$
In actual fact we are told the weight (volume) factor is 2. The difference between 2 and 2·061… is probably within the tolerances we are given in the question, so they probably are similar.

7 The volume factor is $1{\cdot}3^3 = 2{\cdot}197$.
If you could buy them, 2·197 of the small bottles would cost $2{\cdot}197 \times £1{\cdot}40 = £3{\cdot}0758 = £3{\cdot}08$ (to nearest penny).
This would cost more than the £2·40 cost of the large size, so the large size gives you more for your money.

8 The volume scale factor is $\frac{50}{25} = 2$.
So the length scale factor $= \sqrt[3]{2} = 1{\cdot}2599\ldots$
So the large jug is $15 \times 1{\cdot}2599\ldots$ cm high
$= 18{\cdot}898\ldots \text{cm} = 19 \text{cm}$ (to 2 s.f.).

More help or practice
Similarity ► Book Y1 pages 81 to 85
Effects of enlargement on length, area and volume ► Book Y4 pages 9 to 19

Units and measures (page 62)

1 (a) The nearest 5 or 10 grams
(b) The nearest minute
(c) The nearest metre
(d) The nearest 0·1 millimetre
The accuracy will depend to some extent on what the information is needed for. For example, in (b) a TV producer might need the information to the nearest second.

2 (a) 1 gallon is about 4·5 litres; 750 cl = 7·5 litres.
So the order is
3 litres, 1 gallon, 750 centilitres.
(b) 8 ounces $= \frac{1}{2}$ pound $\approx 0{\cdot}25$ kg;
2 pounds ≈ 1 kg.
So the order is 8 ounces, $\frac{1}{2}$ kg, 2 pounds.
(c) 1 inch = 25 mm; 0·01 m = 10 mm.
So the order is 0·01 m, 1 inch, 50 mm.
(d) 1400 metres = 1·4 km; 1 mile $\approx$ 1·6 km.
So the order is
1400 metres, 1 mile, 2 km.

3 12 inches $\approx$ 30 cm, so the circumference
$\approx \pi \times 30 \text{cm} \approx 94 \text{cm}$.
In 1 minute, the record rotates 78 times, so a point on the circumference travels about
$94 \times 78 \text{cm} \approx 7300 \text{cm}$.
So speed is 7300 cm per minute
$= 7300 \div 60$ cm per second
≈ 120 cm per second.

4 26 000 calls in 8 hours = 3250 calls per hour
$= 3250 \div 60$ calls per minute
$= 54{\cdot}16\ldots \approx 50$ calls per minute.

5 *In the following questions you are asked to find the minimum or maximum values of quantities. As these are theoretical values, you do not need to round your answers.*
(a) 3·255 g
(b) The greatest mass of one coin is 3·265 g.
A £5 bag holds 100 5p coins, so the greatest mass of the £5 bag = 326·5 g.
(c) The minimum weight of a £5 bag is
$100 \times 3{\cdot}255 \text{g} = 325{\cdot}5 \text{g}$.
This differs from the maximum weight by exactly 1 g, less than the weight of 1 coin, so the method is valid.

6 Minimum radius = 4·65 cm, so
minimum area = $\pi \times 4{\cdot}65^2 = 67{\cdot}9290\ldots \text{cm}^2$

7 (a) The maximum dimensions of the cuboid are 6·85 cm and 18·45 cm.
So its maximum volume is
$6{\cdot}85 \times 6{\cdot}85 \times 18{\cdot}45 \text{cm}^3 = 865{\cdot}72\ldots \text{cm}^3$
$= 0{\cdot}86572\ldots$ litres
So you might need $5000 \times 0{\cdot}86572\ldots$ litres of juice = 4328·6... litres.

(b) Minimum volume of the carton
$= 6{\cdot}75 \times 6{\cdot}75 \times 18{\cdot}35 \text{cm}^3$
$= 836{\cdot}07\ldots \text{cm}^3 = 0{\cdot}83607\ldots$ litres.
Volume of juice above would fill
$4328{\cdot}6\ldots \div 0{\cdot}83607\ldots$ of these cartons
= 5177·3... = 5177 cartons.

8 Minimum height = 27·5 m
Minimum diameter = 75 m, so
minimum radius = 37·5 m.
So minimum volume
$= \frac{1}{3}\pi \times 37{\cdot}5^2 \times 27{\cdot}5 \text{m}^3 = 40497{\cdot}0\ldots \text{m}^3$.

Maximum height = 32·5 m
Maximum diameter = 85 m, so
maximum radius = 42·5 m.
So maximum volume
$= \frac{1}{3}\pi \times 42{\cdot}5^2 \times 32{\cdot}5 \text{m}^3 = 61473{\cdot}7\ldots \text{m}^3$.

More help or practice

Interval approximations/upper and lower bounds
► Book Y2 pages 12 to 16, Book YX1 pages 40 to 44
Effects of rounding on calculations ► Book Y2 pages 19 to 20
Rates ► Book Y2 pages 34 to 42

Loci (page 64)

All answers are drawn half-size.

1

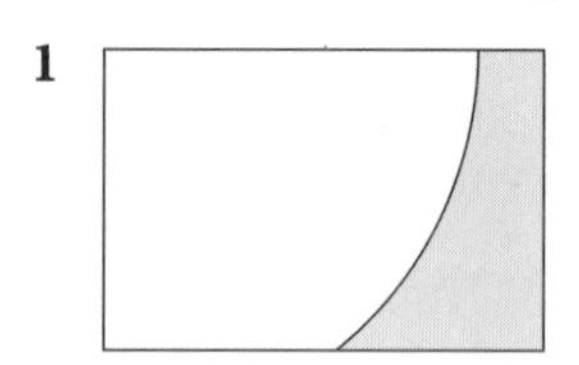

2

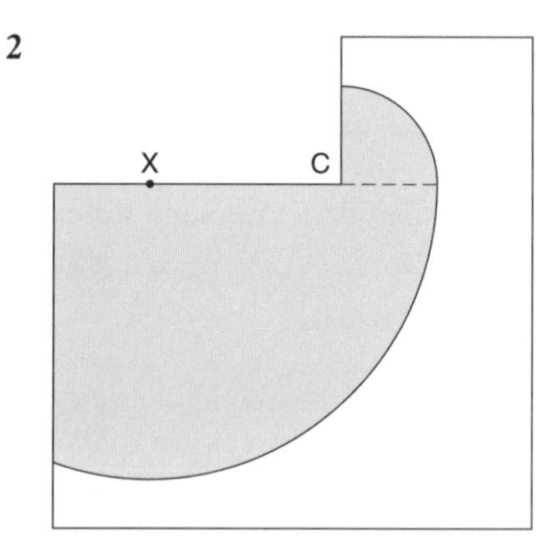

Note that the locus is in two parts. The first part is the inside of a circle centre X, radius 3 m. When the vacuum cleaner flex hits the corner C, the remainder of the locus is the inside of a circle, centre C, with radius 1 m (3 metres minus XC).

3 *The solution below assumes that the power cable continues in a straight line beyond the rectangle shown.*

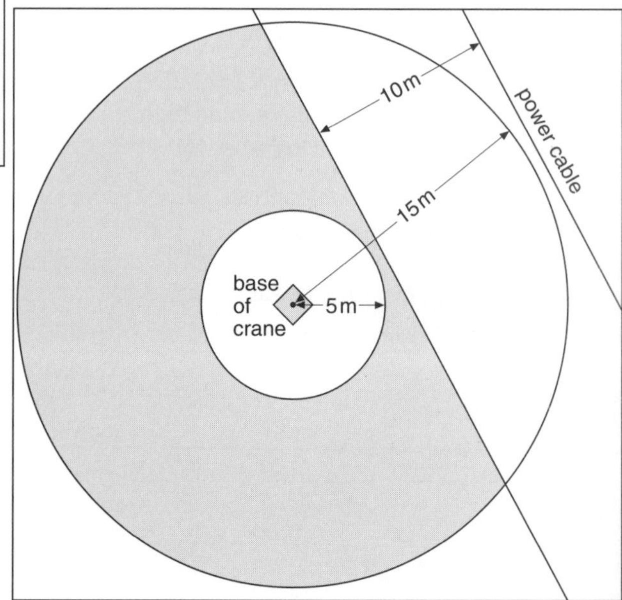

4 (a) and (b)

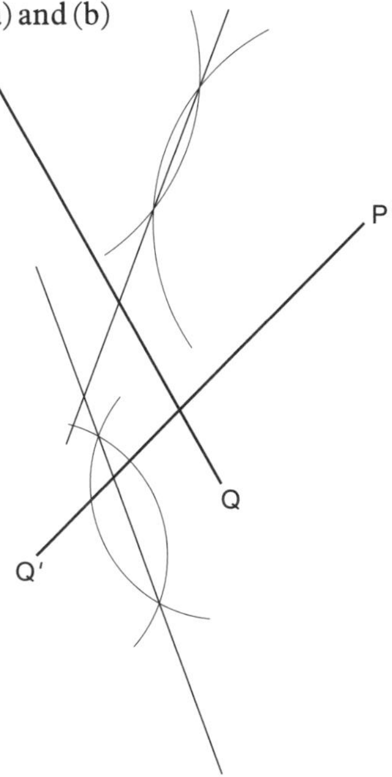

(c) The point is the centre of the rotation that takes the line PQ onto P'Q'.

5

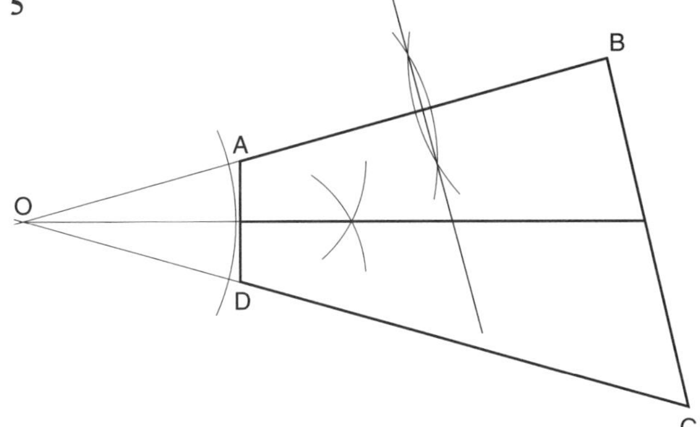

(a) *You first need to extend BA and CD to meet at O.* The electric fence is the bisector of angle AOD.

(b) T is the intersection of the fence with the perpendicular bisector of AB.

More help or practice

Loci ► Book Y1 pages 10 to 15

Trigonometric functions (page 65)

1 Using a calculator, we get
$x = {}^{-}36{\cdot}869\ldots^\circ = {}^{-}37^\circ$ (to the nearest degree).
Now we sketch the graph of $\sin x$.

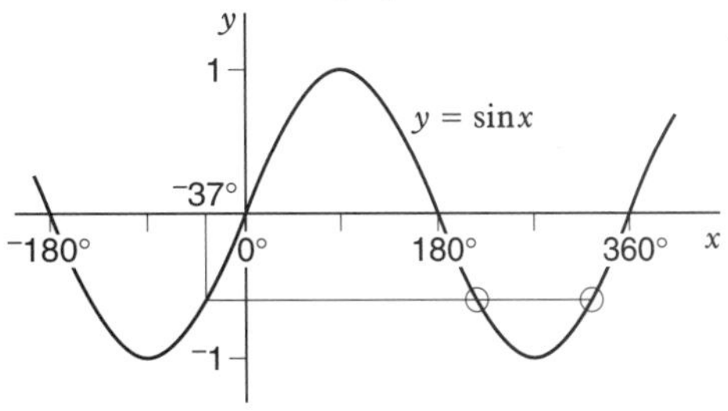

The solutions between 0° and 360° must therefore be $180^\circ + 37^\circ = 217^\circ$ and $360^\circ - 37^\circ = 323^\circ$ (to the nearest degree).
You should now check sin 217° and sin 323° with your calculator.

2 (a)

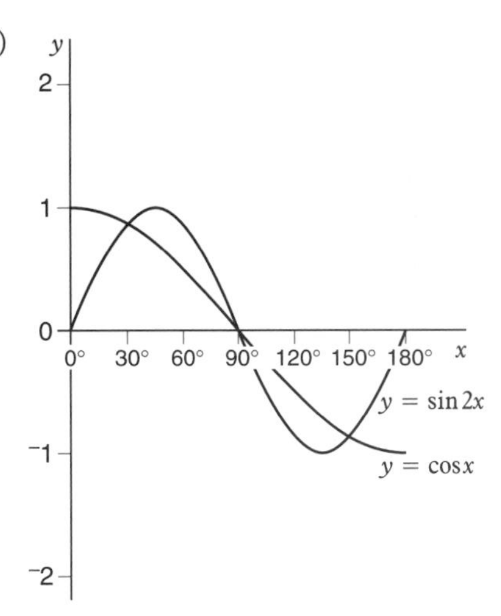

(b) The solutions are where the two graphs cross, that is at about 30°, 90° and 150°.
Now check that $\sin(2 \times 30^\circ) = \cos 30^\circ$, and so on.

3 (a)

(b) $\sin 30° = 0{\cdot}5$, so one solution of $\sin 4x = 0{\cdot}5$ is given by $4x = 30°$, so $x = 7{\cdot}5°$. Also $\sin(180° - 30°) = 0{\cdot}5$, so another solution is given by $4x = 150°$, or $x = 37{\cdot}5°$.

or

$\sin 4x = 0{\cdot}5$ where $2\sin 4x = 1$.

So the solutions are given where $y = 1$ meets the graph of $y = 2\sin 4x$. You can draw $y = 1$ and read off the answers.

4 (a)

t	0	1	2	3	4	5	6	7	8	9	10	11	12
$\cos 30t$	1	0·866	0·5	0	⁻0·5	⁻0·866	⁻1	⁻0·866	⁻0·5	0	0·5	0·866	1
h	4	3·46	2	0	⁻2	⁻3·46	⁻4	⁻3·46	⁻2	0	2	3·46	4

(b)

(c) From the graph, it is possible up to roughly $2\frac{1}{4}$ hours after high water and again after $9\frac{3}{4}$ hours after high water.

Alternatively you could calculate this quite easily.
When $4\cos 30t = 1{\cdot}5$, $\cos 30t = 0{\cdot}375$ and $30t = \cos^{-1}(0{\cdot}375) = 67{\cdot}9756\ldots$, so $t = 2{\cdot}265\ldots$
So it is possible up to 2·27 hours after high water and again after 9·73 hours after high water.

5 (a)

(b)

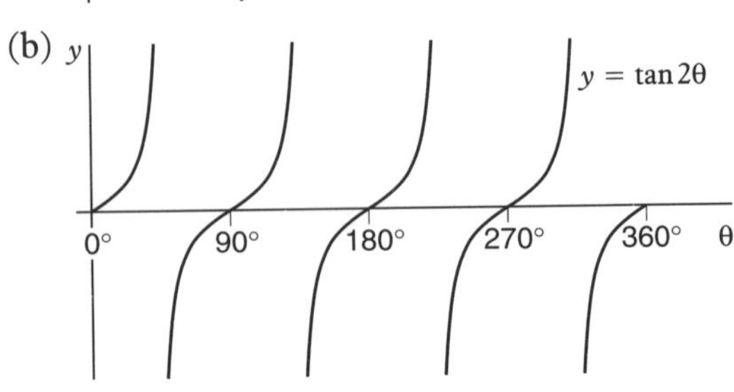

(c) $\tan 2\theta = 1$, so $2\theta + = 45°$ and $\theta = 22{\cdot}5°$.
From the graph of $\tan 2\theta$, we can see that there must be other solutions at $90° + 22{\cdot}5°$, $180° + 22{\cdot}5°$, and $270° + 22{\cdot}5°$.
So the solutions are $22{\cdot}5°$, $112{\cdot}5°$, $202{\cdot}5°$, and $292{\cdot}5°$.
You should check all four of these solutions with your calculator.

6 (a) (i)

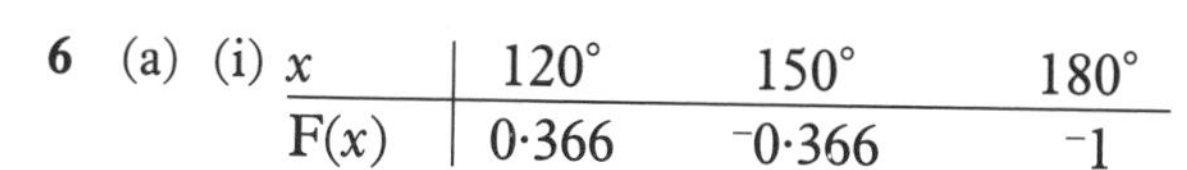

x	120°	150°	180°
F(x)	0·366	⁻0·366	⁻1

(ii)

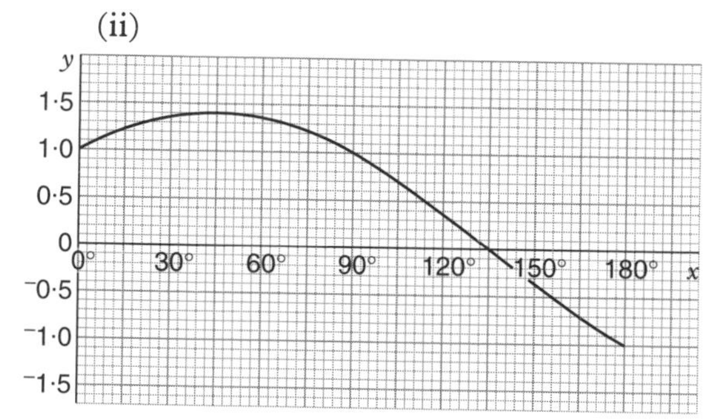

(iii) From the graph, the solution is 135°.
It would be a good idea to check this with your calculator.

(b)

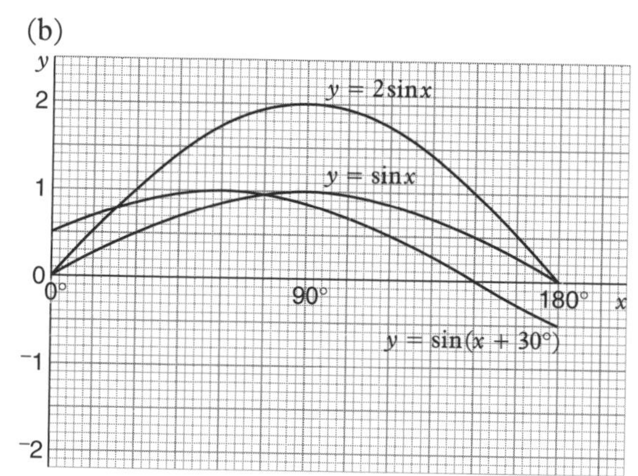

(c) From part (a) we can see that F(x) has an amplitude of 1·4, and its maximum value is at 45° instead of 90°.
So its equation must be
$F(x) = 1{\cdot}4\sin(x + 45°)$.
Hence $p = 1{\cdot}4$ and $q = 45°$.
It would be wise to check this by substituting one or two values for x and seeing if they tally with the graph.

More help or practice

The sine and cosine functions and their graphs ► Book Y5 pages 49 to 53
Inverse sines and cosines ► Book Y5 pages 53 to 56
The tangent function and its graph ► Book Y5 pages 57 to 58
Inverse tangents ► Book Y5 page 59
Graphs and applications of trigonometric functions ► Book Y5 pages 116 to 120

Vectors (page 66)

1 (a) B to C is $\begin{bmatrix} 2 \\ 2 \end{bmatrix}$, C to D is $\begin{bmatrix} 0 \\ 2 \end{bmatrix}$ and D to E is $\begin{bmatrix} 5 \\ 1 \end{bmatrix}$.

(b) $\begin{bmatrix} ^{-}3 \\ 2 \end{bmatrix}$

2 *You might find it helpful to make a sketch to help work out the coordinates.*

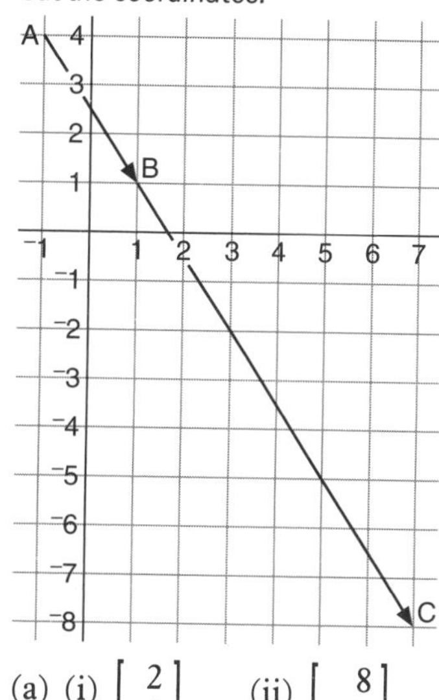

(a) (i) $\begin{bmatrix} 2 \\ ^{-}3 \end{bmatrix}$ (ii) $\begin{bmatrix} 8 \\ ^{-}12 \end{bmatrix}$

(b) Clearly $\overrightarrow{AC} = 4\overrightarrow{AB}$, so they are parallel vectors. Since they both start at A, we deduce that A, B and C lie in a straight line, and that B is $\frac{1}{4}$ of the way from A to C.

3 (a) It is a rhombus.
Parallelogram would be correct, but rhombus is more precise.

(b) (i) $\underset{\sim}{a}$ (ii) $^{-}\underset{\sim}{b}$ (iii) $2\underset{\sim}{a}$
(iv) $\underset{\sim}{b} - \underset{\sim}{a}$ (v) $\underset{\sim}{a} + \underset{\sim}{b}$ (or $\underset{\sim}{b} + \underset{\sim}{a}$)

4 (a) (i) $\overrightarrow{AB} = \overrightarrow{AO} + \overrightarrow{OB} = {}^{-}\overrightarrow{OA} + \overrightarrow{OB}$
$= {}^{-}2\underset{\sim}{a} + (3\underset{\sim}{a} - 2\underset{\sim}{b}) = \underset{\sim}{a} - 2\underset{\sim}{b}$

(ii) $\overrightarrow{BC} = \overrightarrow{OC} - \overrightarrow{OB} = (5\underset{\sim}{a} - 6\underset{\sim}{b}) - (3\underset{\sim}{a} - 2\underset{\sim}{b})$
$= 5\underset{\sim}{a} - 6\underset{\sim}{b} - 3\underset{\sim}{a} + 2\underset{\sim}{b} = 2\underset{\sim}{a} - 4\underset{\sim}{b}$

(b) $\overrightarrow{BC}$ is parallel to $\overrightarrow{AB}$ (and is twice the length of $\overrightarrow{AB}$). But the point B is common to both vectors, so A, B and C must lie on a straight line.

5 (a)

Scale: 1 cm represents 2 N

Resultant

(b) (i) By measurement, the magnitude (length) of the resultant is 17·4 or 17·5.

The resultant of two vectors is simply their sum.

(ii) By measurement, the angle is 20° or 21°.

6 (a) (i) $\overrightarrow{OB'} = 2\overrightarrow{OB} = 2\underset{\sim}{b}$

(ii) $\overrightarrow{AA'} = \overrightarrow{OA} = \underset{\sim}{a}$

(iii) $\overrightarrow{AB} = \overrightarrow{OB} - \overrightarrow{OA} = \underset{\sim}{b} - \underset{\sim}{a}$

(iv) $\overrightarrow{B'A'} = 2\overrightarrow{BA} = 2(\underset{\sim}{a} - \underset{\sim}{b})$ or $2\underset{\sim}{a} - 2\underset{\sim}{b}$

(b) $\overrightarrow{CA'} = \overrightarrow{OA'} - \overrightarrow{OC} = 2\underset{\sim}{a} - (\underset{\sim}{a} + \underset{\sim}{b})$

$= 2\underset{\sim}{a} - \underset{\sim}{a} - \underset{\sim}{b} = \underset{\sim}{a} - \underset{\sim}{b}$

(c) Clearly $\overrightarrow{B'A'} = 2\overrightarrow{CA'}$.

(d) Since both $\overrightarrow{B'A'}$ and $\overrightarrow{CA'}$ end at A′, and $\overrightarrow{B'A'} = 2\overrightarrow{CA'}$, then B′, C and A′ are in a straight line, and C must be the mid-point of B′A′.

More help or practice

Adding and subtracting vectors ► Book Y4 pages 40 to 41, Book YX2 pages 117 to 118

Multiples of a vector ► Book Y4 page 42

Linear sum of vectors ► Book Y4 pages 43 to 47

Vectors in terms of other vectors ► Book YX2 pages 119 to 121

Application of vectors ► Book YX2 pages 126 to 132, 134

Mixed shape, space and measures (page 68)

1 (a) The locus is the arc of a circle, centre R and radius RS.

(b) $TB = TR + RB = (4{\cdot}8\tan 50° + 0{\cdot}5)\,m$
$= (5{\cdot}720\ldots + 0{\cdot}5)\,m$
$= 6{\cdot}2\,m$ (to 2 s.f.)

(c) ΔPBQ is right-angled at B and
$PB = (6{\cdot}220\ldots - 1{\cdot}2)\,m$
$= 5{\cdot}020\ldots\,m$

$\cos P = \frac{5{\cdot}020\ldots}{8{\cdot}5} = 0{\cdot}59063\ldots$ and

$P = 53{\cdot}797\ldots°$
$= 54°$ (to the nearest degree)

2 (a) Volume of pyramid $= \frac{1}{3} \times 230^2 \times 146{\cdot}5\,m^3$
$= 2583283{\cdot}3\ldots m^3 = 2{\cdot}5832\ldots \times 10^6\,m^3$
Mass $= 2{\cdot}8 \times$ volume $= 7{\cdot}2331\ldots \times 10^6\,kg$
$= 7{\cdot}23 \times 10^6\,kg$ (to 3 s.f.)

(b)

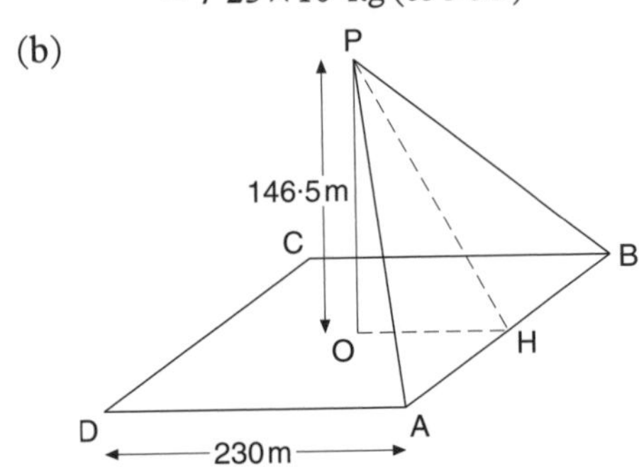

A sketch of one face of the pyramid is as above. Clearly OH = 115 m.
So $PH = \sqrt{(146{\cdot}5^2 + 115^2)}\,m = 186{\cdot}2\ldots m$.
Area of PAB $= \frac{1}{2} \times PH \times AB$
$= \frac{1}{2} \times 186{\cdot}2\ldots \times 230\,m^2$

$= 2{\cdot}141\ldots \times 10^4\,m^2$
So area of all 4 faces $= 8{\cdot}567\ldots \times 10^4\,m^2$
$= 8{\cdot}57 \times 10^4\,m^2$ (to 3 s.f.)

Note you could give your answer as 85 700 m², since this part does not ask for standard form.

3 (a) (i) $2\underset{\sim}{b} - 2\underset{\sim}{a}$ (ii) $4\underset{\sim}{b}$ (iii) $3\underset{\sim}{a} + 4\underset{\sim}{b}$

(b) Length of $\underset{\sim}{a} = \sqrt{(3^2 + 4^2)} = \sqrt{25} = 5$ units

(c) (i) 10 units (ii) 20 units (iii) 15 units

(d) Angle in a semicircle

(e) (i) 13·2 (ii) 41·4° (iii) 20·7°

4 (a) *The angle at the centre of a regular pentagon is $360^\circ \div 5 = 72^\circ$.*

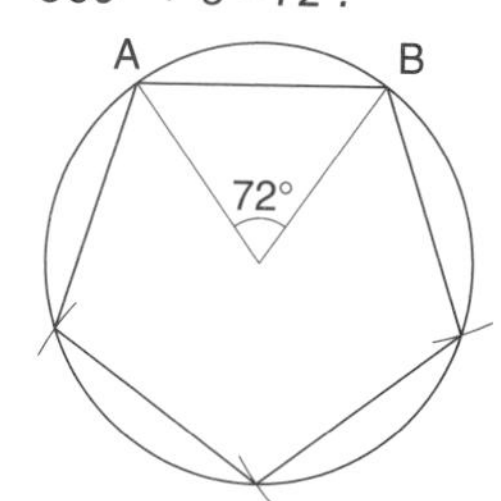

Draw a circle of radius 6 cm and draw two radii at an exact angle of 72°. Then use compasses to mark off three more arcs equal to AB around the circle. This method is probably more accurate than the alternative method of measuring three more angles of 72°. *Whichever method you use, either check whether the fifth arc is equal to the other four, or whether the fifth angle is 72°.*

(b)

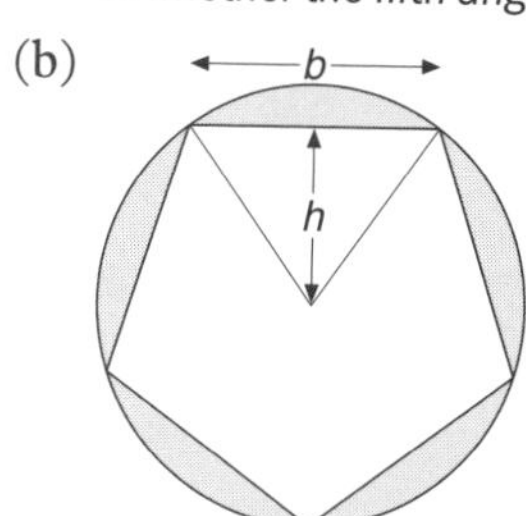

Measure the distances b and h marked in the diagram and hence find the area of the pentagon which is given by

$5 \times \frac{1}{2} \times b \times h$.

Acceptable accuracies are as follows:

h between 4·7 cm and 5·0 cm
b between 6·9 cm and 7·2 cm
Area of pentagon between 80 cm² and 90 cm²

The area of the circle $= (\pi \times 6^2)\text{cm}^2 \approx 113\,\text{cm}^2$
So the area of the five shaded segments is the difference between the two.
An answer between 23 cm² and 33 cm² is acceptable.

5 (a) *It often helps to mark the distances and angles you know on a sketch.*
The height of the building in metres is given by $83{\cdot}4 \tan 32^\circ + 1{\cdot}6 = 53{\cdot}714\ldots$
$= 53{\cdot}7$ (to 3 s.f.).

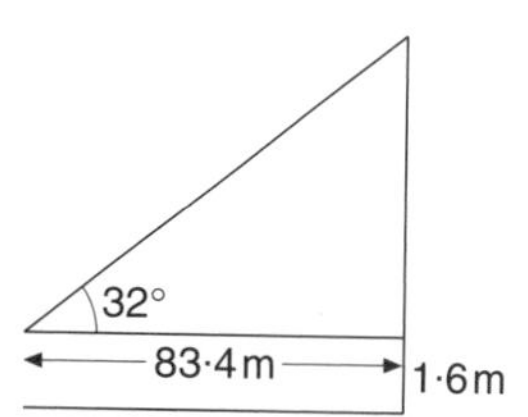

(b) The maximum possible height in metres is given by
$83{\cdot}45 \tan 32{\cdot}5^\circ + 1{\cdot}6 = 54{\cdot}763\ldots$
$= 54{\cdot}8$ (to 3 s.f.).

6

(a) In the diagram $\text{AM} = (90 - 40)\text{mm} = 50\,\text{mm}$
$\text{AR}^2 = \text{MR}^2 - \text{AM}^2 = 500^2 - 50^2 = 247\,500$
$\text{AR} = 497{\cdot}493\ldots\,\text{mm}$
$\text{PQ} - \text{AR} = (500 - 497{\cdot}493\ldots)\text{mm} \approx 2{\cdot}5\,\text{mm}$

(b) Area $= \frac{1}{2}(90 + 40) \times 497{\cdot}5\,\text{mm}^2$
$= 32\,337{\cdot}5\,\text{mm}^2$
$= 32\,300\,\text{mm}^2$ (to 3 s.f.)

(c) $\cos \text{APQ} = \cos \text{AMR} = \frac{50}{500}$ and
$\angle\text{APQ} = 84{\cdot}260\ldots^\circ$
$\approx 84{\cdot}3^\circ$

(d) Area of (sector ACB + sector RST + 2 × ARQP)
$= \frac{360 - 2 \times 84{\cdot}3}{360} \times \pi \times 90^2$
$+ \frac{2 \times 84{\cdot}3}{360} \times \pi \times 40^2 + 2 \times 32\,337{\cdot}5)\,\text{mm}^2$
$= (13\,529{\cdot}2\ldots + 2354{\cdot}10\ldots + 64\,675)\text{mm}^2$
$= 80\,558{\cdot}3\ldots\,\text{mm}^2 = 80\,600\,\text{mm}^2$ (to 3 s.f.)

It was reasonable to use the rounded values of the angles in a question like this. It would also have been acceptable to use the rounded value 32 300 mm² for the area of ARQP. This would give an answer of 80 500 mm² (to 3 s.f.).

7 (a) BE is a diagonal of the square BDEF, so
$BE^2 = BD^2 + DE^2 = 2 \times 0{\cdot}8^2$
$BE = \sqrt{1{\cdot}28}\,\text{mm} = 1{\cdot}131\ldots\text{mm}$
$= 1{\cdot}1\,\text{mm}$ (to 2 s.f.)

(b) The angle is shown in the diagram.

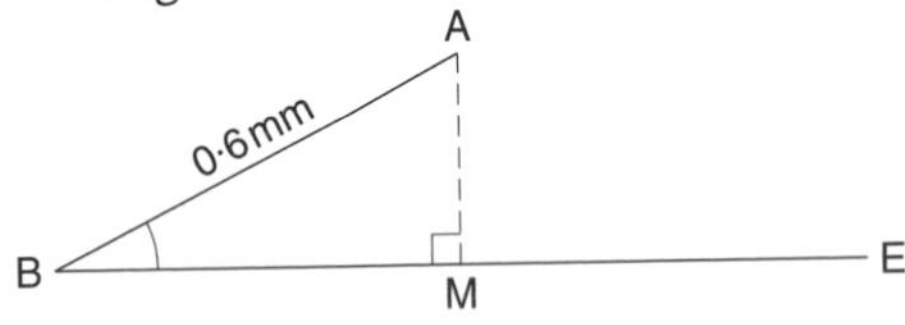

$BM = \frac{1}{2}BE$ and

$$\cos ABM = \frac{BM}{BA} = \frac{1{\cdot}131\ldots \div 2}{0{\cdot}6} = 0{\cdot}9428\ldots$$

$\angle ABM = 19{\cdot}471\ldots^\circ$
$= 19^\circ$ (to the nearest degree)

(c) *By symmetry the height AC is twice AM.*
$AM = AB \sin ABM$
$= 0{\cdot}6 \sin(19{\cdot}471\ldots) = 0{\cdot}2\,\text{mm}$
$AC = 2 \times 0{\cdot}2\,\text{mm} = 0{\cdot}4\,\text{mm}$
You could also have answered this by using Pythagoras' rule.

(d) Volume of diamond
$= 2 \times$ volume of pyramid ABDEF
$= 2 \times \frac{1}{3} \times \text{base area} \times \text{height}$
$= 2 \times \frac{1}{3} \times 0{\cdot}8^2 \times 0{\cdot}2\,\text{mm}^3$
$= 0{\cdot}0853\ldots\text{mm}^3$
$= 0{\cdot}085\ldots\text{mm}^3$ (to 2 s.f.)

8 (a)

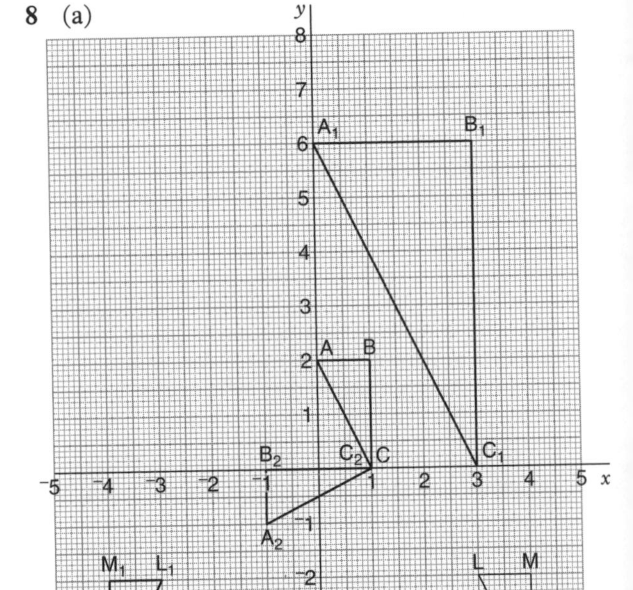

(b) Translation $\begin{bmatrix} 3 \\ ^-4 \end{bmatrix}$

(c) Reflection, in line $y = {}^-x$
You need to fully describe the line of reflection to gain full marks.

9 (a) *The loaf of bread is a prism whose cross-section can be divided into a rectangle and semi-circle. The volume of a prism is given by area of cross-section × length.*

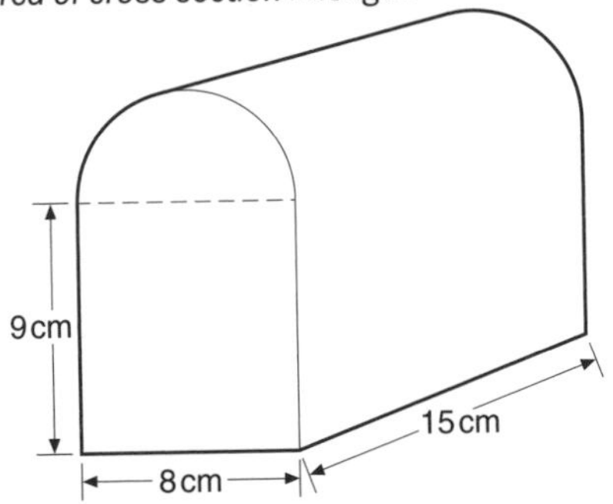

Volume $= [(9 \times 8) + (\frac{1}{2} \times \pi \times 4^2)] \times 15\,\text{cm}^3$
$= 1456{\cdot}991\ldots\text{cm}^3$
$= 1457\,\text{cm}^3$ (to nearest cm^3)

(b) (i) Volume of balloon

$= 1456{\cdot}991\ldots \times 50^3 \text{ cm}^3$
$= (1{\cdot}82123\ldots \times 10^8) \times 10^{-6} \text{m}^3$
$= 182 \text{m}^3$ (to nearest m^3)

*The volume scale factor is the **cube** of the linear scale factor.*

(ii) *The area scale factor is the **square** of the linear scale factor.*

Surface area

$= [(2 \times \text{area of cross-section}) + (\text{area of rectangular base}) + (\text{area of curved surface})] \times 50^2 \text{cm}^2$
$= \{2[(9 \times 8) + (\frac{1}{2}\pi \times 4^2)] + (8 \times 15) + 15(4\pi + 18)\} \times 2500 \times 10^{-4} \text{m}^2$
$= 193{\cdot}190\ldots \text{m}^2$
$= 193 \text{m}^2$ (to nearest m^2)

HANDLING DATA

Pie charts (page 71)

When you draw a pie chart in answer to a question, marks will almost certainly be awarded for labelling each sector.

1 (a)

	£ (billions)	%
Health and Social Security	56·5	42·8
Defence	18·1	13·7
Education	14·2	10·8
Housing and Environment	5·4	4·1
Industry and Transport	8·5	6·4
Other	29·3	22·2
Total	132·0	100·0

(b)

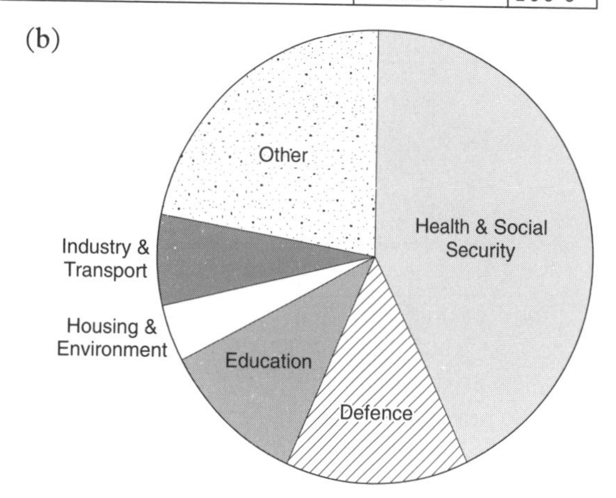

2 (a) 48% of the petroleum was used by cars and taxis.

(b) 20% was used by ships and aircraft altogether.

(c) 14 million tonnes were used by ships and aircraft altogether.

More help or practice

Pie charts ► Book R1 pages 122 to 123

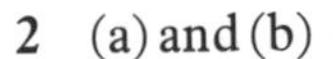

Scatter diagrams (page 72)

Normally in an exam you will be given the grid on which to plot the scatter diagrams. The normal convention is for the axis for the first set of data to be along the bottom of the grid and the axis for the second set to be up the side of the grid. If your diagrams look different from those shown here, first check that you have kept to this convention.

1

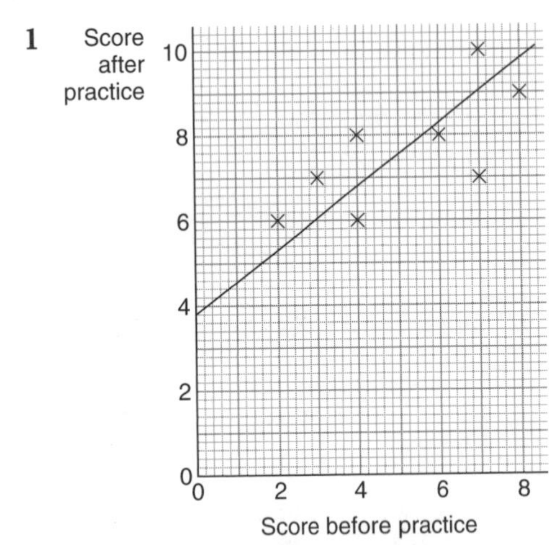

You could work out the means to help you draw your line of best fit but you do not have to unless the question tells you to.

The practice seems to have improved pupils' scores, though those who were poorer with their tables to start with tended to improve more than those who were better to start with. One pupil's score stayed the same. In this case the moderate correlation shown by the scatter diagram tells us that the better pupils generally stayed better – in other words, not much 'overtaking' occurred.

2 (a) and (b)

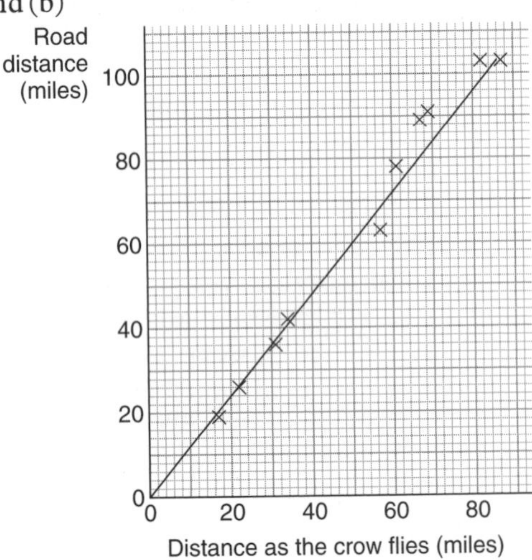

(c) *Work out the gradient of the line to find a rule of thumb.*

A rough rule is 'Take the "as the crow flies distance" and add on 20% (one-fifth)' or 'Multiply the "as the crow flies distance" by 1·2'. Your rule might be slightly different from this.

3 (a) (i) The mean of the maths scores is 37·25.

(ii) The mean of the French scores is 23·25.

(b) (i) The crosses show the scores.

(ii) The dot shows the mean scores.

(iii) The line shown is the line of best fit.

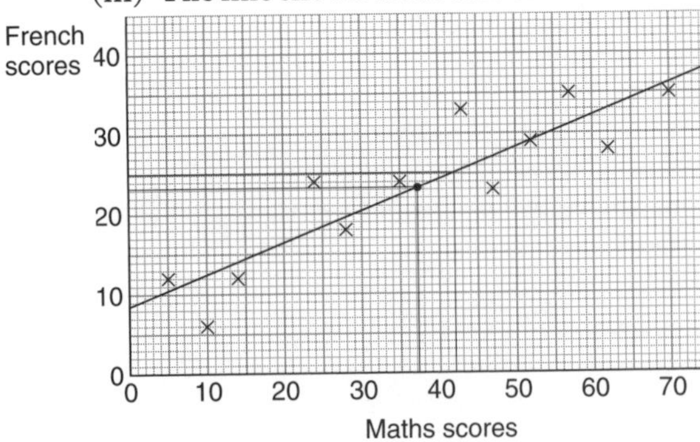

(c) The line of best fit suggests a French score of 25.

(d) Other pupils' French scores were as much as 8 marks above or below the line. So we cannot be sure that she would have got a score of 25.

> **More help or practice**
> Scatter diagrams ► Book Y4 pages 120 to 121
> Correlation ► *Kids like us* pages 10 to 11

Mode, mean, median and range (page 74)

1 (a) Leslie's total score = 632
Leslie's mean score = 632 ÷ 8 = 79

(b) His range is 91 – 73 = 18.

(c) You might choose Pat because he is more consistent (has the lower range) or because he has fewer lower scores.
You might choose Leslie because, although his mean is the same as Pat's, he has the highest score.
Any one of these reasons would gain you the marks.

2 (a)

Mark	0	1	2	3	4	5	6	7	8	9	10
Number of boys	1	1	0	2	4	3	5	8	3	1	2

Check that the number of boys totals 30.

(b) Total marks
$= (0\times1) + (1\times1) + (2\times0) + (3\times2) + (4\times4) + \ldots$
$= 0 + 1 + 0 + 6 + 16 + 15 + 30 + 56 + 24 + 9 + 20$
$= 177$

(c) (i) The modal mark is 7.
(ii) The median is the mean of the 15th and 16th marks = (6 + 6) ÷ 2 = 6.
(iii) The mean = 177 ÷ 30 = 5·9

(d) (i) Total marks = 6·5 × 20 = 130
(ii) The overall mean mark
= (177 + 130) ÷ 50
= 6·14

3 (a) The modal number bought is 10 (by 6 families).

(b) The median number is the average of the 10th and 11th numbers (arranged in order)
= (8 + 10) ÷ 2 = 9.
A stem-and-leaf table is often useful for finding the median quickly – see Book Y3, page 87.

(c) The mean number = 168 ÷ 20 = 8·4

4 (a) Amal has scored 10·5 × 8 = 84 goals.

(b) To improve on her mean score she needs to score a total of (11·2 × 9) goals = 101 goals (rounded to the nearest number *above*).
So she needs to score 101 – 84 = 17 goals to improve her mean average.

More help or practice

Mean of ungrouped and grouped data ► Book Y2 pages 70 to 71
Median ► Book Y3 pages 84 to 88
Range ► Book Y3 page 89
Comparing averages ► Book YR+ pages 20 to 25, Book Y3 pages 89 to 90

Cumulative frequency (page 76)

1 (a)

Attendance 1000's	Frequency	Cumulative frequency
0–5000	15	15
5001–10000	12	27
10001–15000	2	29
15001–20000	5	34
20001–25000	2	36
25001–30000	2	38
30001–35000	2	40

(b)

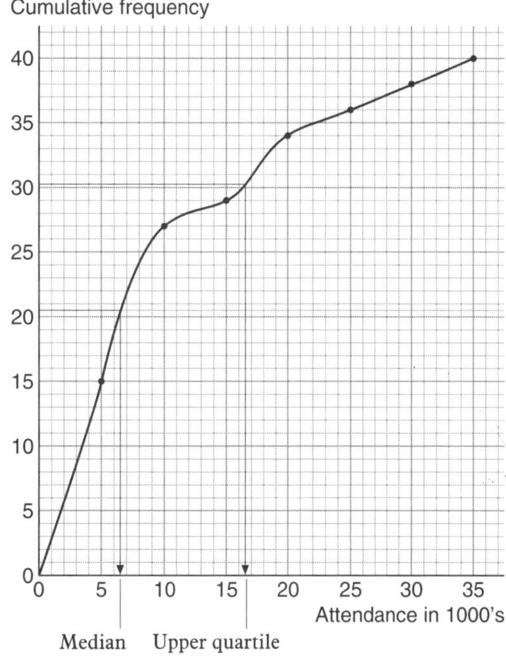

(c) (i) The median attendance is 6500.

An answer between 6000 and 7000 is acceptable.

(ii) The upper quartile is 16 600.

An answer between 16 000 and 17 000 is acceptable.

2 (a)

Length l (in minutes)	Number of calls	Cumulative frequency
$0 < l \le 4$	24	24
$4 < l \le 8$	47	71
$8 < l \le 12$	68	139
$12 < l \le 16$	44	183
$16 < l \le 20$	32	215
$20 < l \le 24$	21	236
$24 < l \le 28$	8	244
$28 < l \le 32$	6	250

(b)

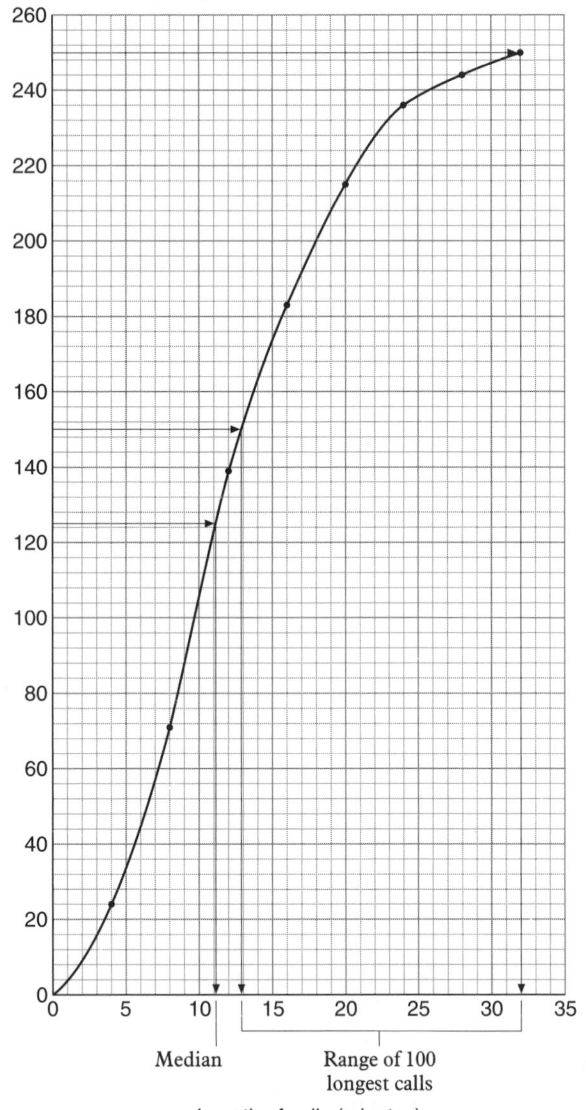

(c) (i) The median length of call is 11 minutes.

(ii) The range of the hundred longest calls is 19 minutes.

Answers of between 10 and 12 minutes and between 18 and 20 minutes, respectively, are acceptable.

3 (a)

Waist-to-knee length w (cm)	Cumulative frequency
$32\text{cm} < w \leq 34\text{cm}$	3
$32\text{cm} < w \leq 36\text{cm}$	19
$32\text{cm} < w \leq 38\text{cm}$	66
$32\text{cm} < w \leq 40\text{cm}$	91
$32\text{cm} < w \leq 42\text{cm}$	96
$32\text{cm} < w \leq 44\text{cm}$	100

(b)

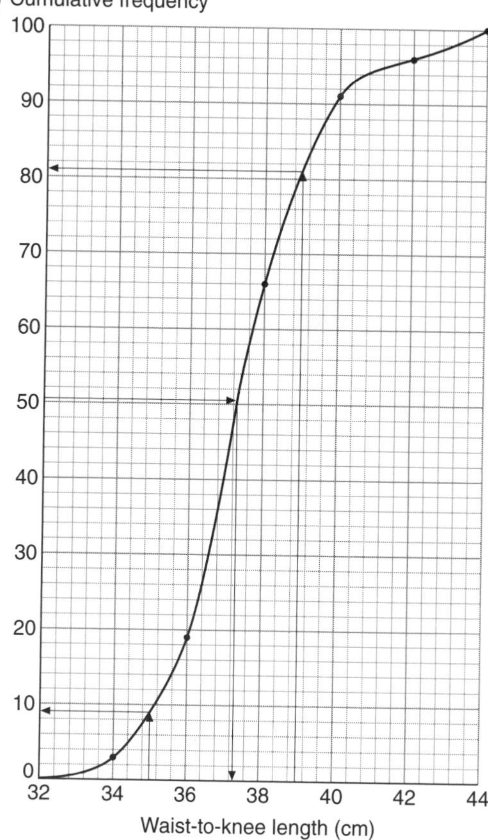

(c) The median is about 37 cm.

(d) 35 cm to 39 cm will fit 72% of the sample (from 9% to 81%).

More help or practice

Percentiles ► Book Y3 pages 140 to 141

Medians and quartiles ► Book Y3 pages 141 to 143

Cumulative frequency graphs ► Book Y3 pages 144 to 147, 149 to 151

Interquartile range ► Book Y3 page 148

Standard deviation (page 78)

1 Mean = 4·6, standard deviation = 0·16

2 (a) *Machine A*
Mean = 454 g, standard deviation = 12 g
Machine B
Mean = 454 g, standard deviation = 3 g

(b) On average the machines both pack the required amount of honey into the jars. But jars from machine A will vary in actual weight more than those from machine B. The variability of machine A would almost certainly be unacceptable commercially.

3 (a) *All these answers are in grams.*
Adam
Mean = 152·3
Standard deviation = $\sqrt{11 \cdot 81}$ = 3·436...
= 3·4 (to 1 d.p.)
Sven
Mean = 151·8
Standard deviation = $\sqrt{3 \cdot 16}$ = 1·777...
= 1·8 (to 1 d.p.)
Cindy
Mean = 149·3
Standard deviation = $\sqrt{4 \cdot 01}$ = 2·002...
= 2·0 (to 1 d.p.)

(b) Cindy achieved a mean weight closest to 150 grams, but her mean of 149·3 grams is under-weight, so the manager would be most pleased with Sven because his mean is the closest (above) to 150 grams and his standard deviation is the lowest.

4 (a) *Station A*
Mean = 838 ÷ 14 = 59·9 (to 1 d.p.)
Standard deviation = 6·7 (to 1 d.p.)
Station B
Mean = 916 ÷ 14 = 65·4
Standard deviation = 2·9 (to 1 d.p.)

(b) Although station A has the lower mean, B has a much lower standard deviation, so there is some chance that more ambulances would be needed at station A on a busy day.

More help or practice

Standard deviation of ungrouped data
► Book Y5 pages 64 to 66, Book YX2 pages 110 to 111

Estimates from grouped frequency data (page 80)

1

Wage (£w)	Frequency	Mid-interval value	Frequency × mid-interval value
$150 \le w < 200$	20	175	3500
$200 \le w < 250$	12	225	2700
$250 \le w < 300$	7	275	1925
$300 \le w < 350$	4	325	1300
$350 \le w < 400$	5	375	1875
$400 \le w < 450$	0	425	0
$450 \le w < 500$	0	475	0
$500 \le w < 550$	0	525	0
$550 \le w < 600$	1	575	575
$600 \le w < 650$	1	625	625
		Total	12500

The estimated mean is £$\frac{12500}{50}$ = £250.

2 *You will find it helps to create extra columns to the table, like this.*

Mark	Frequency (f)	Mid-interval value (X)	fX	fX^2
1–5	1	3	3	9
6–10	1	8	8	64
11–15	5	13	65	845
16–20	10	18	180	3240
21–25	21	23	483	11109
26–30	9	28	252	7056
31–35	4	33	132	4356
36–40	4	38	152	5776
41–45	3	43	129	5547
Totals	58		1404	38002

Mean = 1404 ÷ 58 = 24·206… = 24·2 (to 3 s.f.)

Standard deviation = $\sqrt{\left[\frac{38\,002}{58} - \left(\frac{1404}{58}\right)^2\right]}$

= 8·32 (to 3 s.f.)

3 (a) Mean (in grams) = 4335 ÷ 80 = 54·1875
= 54 (to nearest gram)

(b) (i) 14 eggs weigh more than 60 g.
This is 17·5% of the sample.

(ii) 420 eggs

More help or practice

Estimating the mean of grouped data ► Book Y2 pages 71 to 75, Book Y5 pages 67 to 68

Estimating standard deviation of grouped data ► Book Y5 pages 67 to 68

Histograms (page 82)

1 *Frequency = frequency density × width of interval*

(a) $20 \times 0{\cdot}4 = 8$
8 tuna weighed less than 20 kg.

(b) $10 \times 1{\cdot}2 = 12$
12 tuna weighed between 20 kg and 30 kg.

(c) $8 + 12 + (10 \times 1{\cdot}6) + (20 \times 0{\cdot}2) = 20 + 16 + 4$
A total of 40 tuna were caught.

(d) 10% weighed 40 kg or more.

2 (a)

Age in years, a	Frequency
$12 \le a < 17$	18
$17 \le a < 23$	30
$23 \le a < 35$	12
$35 \le a < 50$	9
$50 \le a < 65$	6
$65 \le a < 80$	3

(b)

Age in years, a	Frequency	Frequency density
$12 \le a < 17$	5	1·0
$17 \le a < 23$	3	0·5
$23 \le a < 35$	18	1·5
$35 \le a < 50$	42	2·8
$50 \le a < 65$	21	1·4
$65 \le a < 80$	16	1·1

(c)

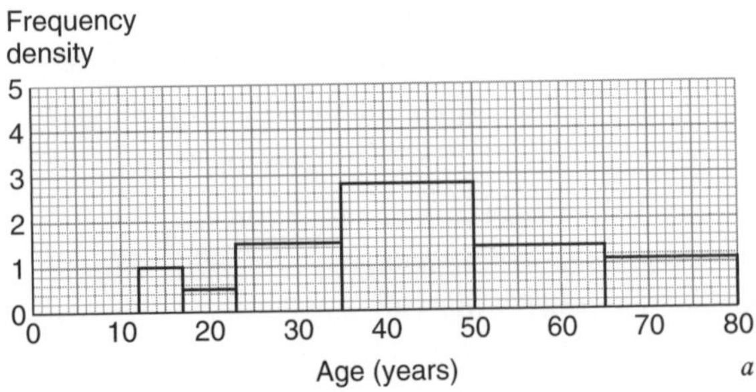

(d) The majority of people interviewed at the record store were teenagers or in their twenties whereas most of the people at the garden centre were over 35 years of age.

3

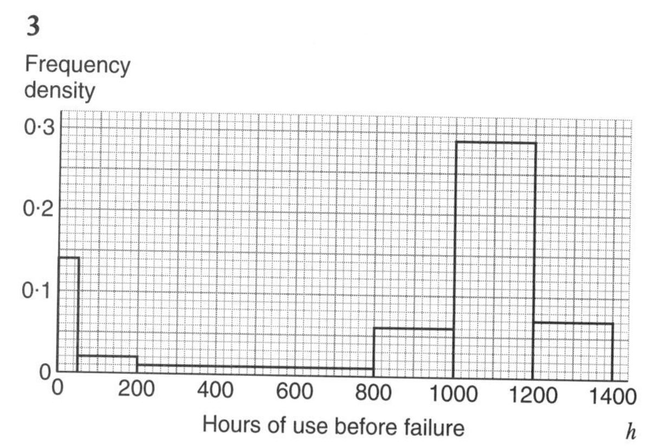

The frequency densities are 0·14, 0·02, 0·01, 0·06, 0·29 and 0·07, respectively.

4 (a)

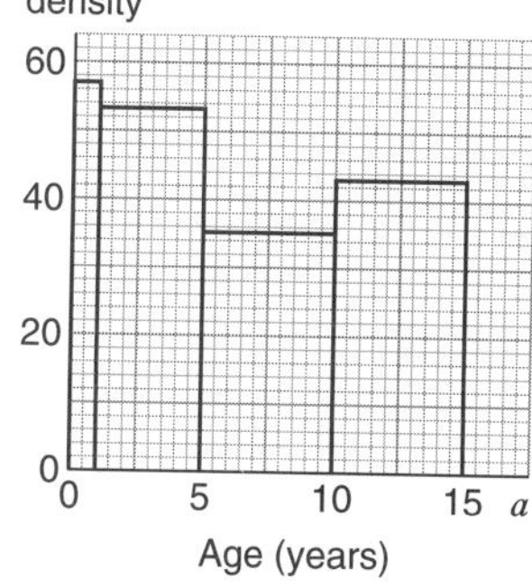

The frequency densities are 57, 53·25, 35·2 and 43, respectively.

(b) Children under 1 year old have the highest risk of accidental death.

Notice that the histogram gives a better indication of the riskiest age range than the table. There were more accidents in the $1 \le a < 5$ age range than in the $0 \le a < 1$ range; but this is because $1 \le a < 5$ is a wider range (with more children of that age in the population as a whole), not because it is riskier.

More help or practice

Frequency diagrams for grouped data
► Book YR+ pages 26 to 28, Book Y2 pages 67 to 72

Frequency polygons ► Book YR+ page 28,
Book YX2 pages 105 to 106

Histograms and frequency density
► Book YX2 pages 107 to 111

Approaches to sampling (page 84)

These are some possible answers to the questions. You may have made some equally valid suggestions for which you would gain the marks.

1 Group according to age: for example include people with young children, car drivers, cyclists and shopkeepers in the correct proportions.

2 *First work out the total number of staff and so find the fraction of the workforce that each group represents. Then work out the size of each sample and round appropriately.*

$\frac{159}{1342} \times 150 = 17{\cdot}771\ldots$ of the sample should be management.

Similarly, the other two calculations are

$\frac{428}{1342} \times 150 = 47{\cdot}839\ldots$ and $\frac{755}{1342} \times 150 = 84{\cdot}388\ldots$

So the size of each sample should be

Management 18,
Secretarial and clerical 48,
Shop floor staff 84.

3 *Your answer might include points like these.*

Using a local phone directory would not give him a sample from the country as a whole.

Some pages of the phone directory have large blocks of common names, such as Patel, Jones or Smith, and if he hit one of these his sample would not be culturally representative.

Interviewing by phone would miss out those young people who lived in bedsits or hostels with no phone of their own.

Some phone calls would be wasted if there was no young person at that number.

4 Some ready meals can be quite expensive. The supermarket might have been in a low-income or high-income area, so the sample would not necessarily represent the range of incomes of people shopping at the chain of supermarkets overall. Some people, like those out at work all day, buy ready meals because they don't have time to cook. But the survey was done when most people like those would be at work. The survey included people who do the shopping, but they are not the only ones who do the eating.

Some people might refuse to be interviewed. These people might have distinctive views about the subject, so the sample would not be representative of the population. This point applies to most kinds of surveys, so it is often worth stating it if you think you have not given enough valid criticisms of a sampling method in your answer (for example, it also applies to questions 3 and 6).

5 Albert High School 11
St Joseph's High School 19
London Road School 20

6 This may be a street where the even numbers are all on the same side and it may turn out that there are rented council houses on the even side but privately owned houses on the odd side. So Katya's sample would not be representative.

Also, she is getting a sample of houses when she should be getting a sample of people. If she just interviews the person who answers the door the sample may not turn out to be representative.

More help or practice

Representative and biased samples
► Book Y4 pages 110 to 111

Random sampling ► Book Y4 pages 113 to 116

Probability (page 86)

*Probabilities can be given as fractions, decimals or percentages but **not** as ratios. For example in question 1, 8:25 would be penalised.*

1 (a) $\frac{8}{25}$ or 0·32 or 32%

(b) $\frac{8+7}{25} = \frac{15}{25} = \frac{3}{5}$ or 0·6 or 60%

2 (a) (i) $\frac{26}{100}$ or $\frac{13}{50}$ or 0·26 or 26%

(ii) She would need to do some more 'spins'.

(b) The probability of a four is $\frac{20}{100} = \frac{1}{5}$.

The probability of a one is $\frac{5}{100} = \frac{1}{20}$.

So the probability of a four followed by a one is $\frac{1}{5} \times \frac{1}{20} = \frac{1}{100}$ or 0·01.

3 (a)

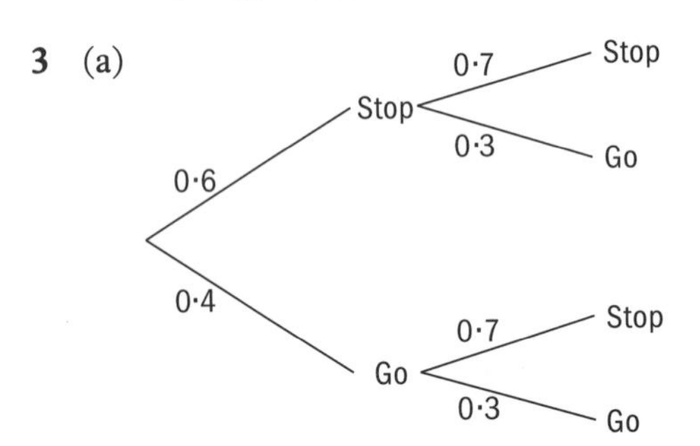

(b) *There are two ways of answering this question.*

***Either** add the three outcomes in which he stops.*

$(0{\cdot}6 \times 0{\cdot}7) + (0{\cdot}6 \times 0{\cdot}3) + (0{\cdot}4 \times 0{\cdot}7)$

$= 0{\cdot}42 + 0{\cdot}18 + 0{\cdot}28$

$= 0{\cdot}88$

***Or** use the fact that the probability that he stops at least once is*

1 – (the probability that he does not stop at all).

$1 - (0{\cdot}4 \times 0{\cdot}3) = 1 - 0{\cdot}12 = 0{\cdot}88$

Remember to show all your working so that you can gain marks for method even if you make a mistake in the calculation.

4 (a) 0·15

(b) (i) $0{\cdot}85 \times 0{\cdot}85 \times 0{\cdot}85 = 0{\cdot}614125$

$= 0{\cdot}61$ (to 2 s.f.)

A probability cannot be bigger than 1; if you got the answer 2·55 you made the common mistake of adding instead of multiplying.

(ii) $1 - 0{\cdot}614125 = 0{\cdot}39$ (to 2 s.f.)

5 (a) $\frac{1}{5} \times \frac{1}{5} \times \frac{1}{5} = \frac{1}{125}$

(b) $\frac{4}{5} \times \frac{4}{5} \times \frac{4}{5} = \frac{64}{125}$

(c) $\frac{1}{125} \times 5 = \frac{1}{25}$

The answer to (a) mulitplied by 5.

(d) $\frac{4}{5} \times \frac{3}{5} = \frac{12}{25}$

Whatever the first day is, there are 4 choices for the second day and 3 choices for the third.

6 (a)

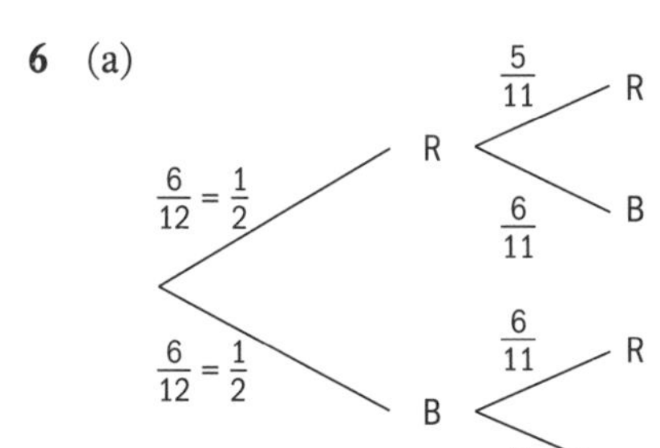

The probability of two reds or two blacks is $\frac{1}{2} \times \frac{5}{11} + \frac{1}{2} \times \frac{5}{11} = \frac{5}{22} + \frac{5}{22} = \frac{5}{11}$.

You might have got this answer by thinking 'whatever colour card he takes first there will be five of that colour out of the 11 that are left'.

(b)

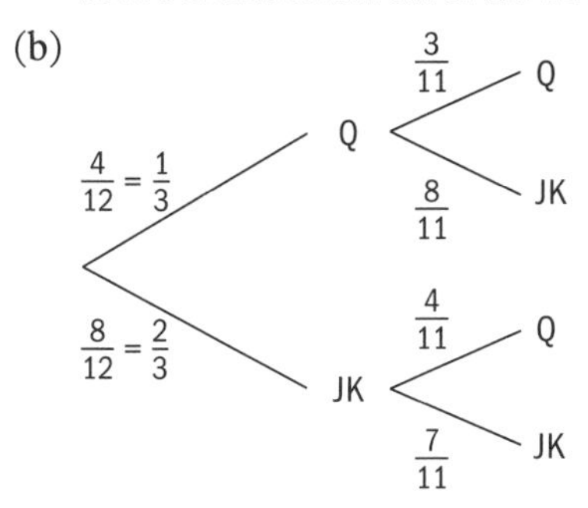

The probability of one queen only is $\frac{1}{3} \times \frac{8}{11} + \frac{2}{3} \times \frac{4}{11} = \frac{8}{33} + \frac{8}{33} = \frac{16}{33}$.

(c) The probability of a red queen and red king is $\frac{2}{12} \times \frac{2}{11} = \frac{4}{132} = \frac{1}{33}$.

There are eight such possible combinations, all equally likely.

So the probability of the required outcome is $\frac{1}{33} \times 8 = \frac{8}{33}$.

7

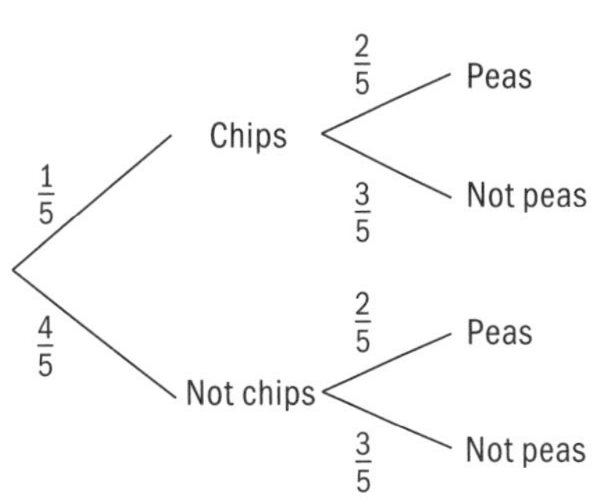

$$\frac{1}{5} \times \frac{3}{5} + \frac{4}{5} \times \frac{2}{5} = \frac{3}{25} + \frac{8}{25}$$
$$= \frac{11}{25} \text{ or } 0{\cdot}44$$

8 (a) $\frac{1}{27}$

(b) The probability of three arrows pointing to the same shape is $\frac{1}{27} + \frac{1}{27} + \frac{1}{27} = \frac{1}{9}$.

So the probability of winning £1 is $\frac{1}{9}$ or $0{\cdot}\dot{1}$.

(c) The money taken is £80.

The money kept is £80 – 400 × $\frac{1}{9}$ × £1 = £35·56.

9 (a)

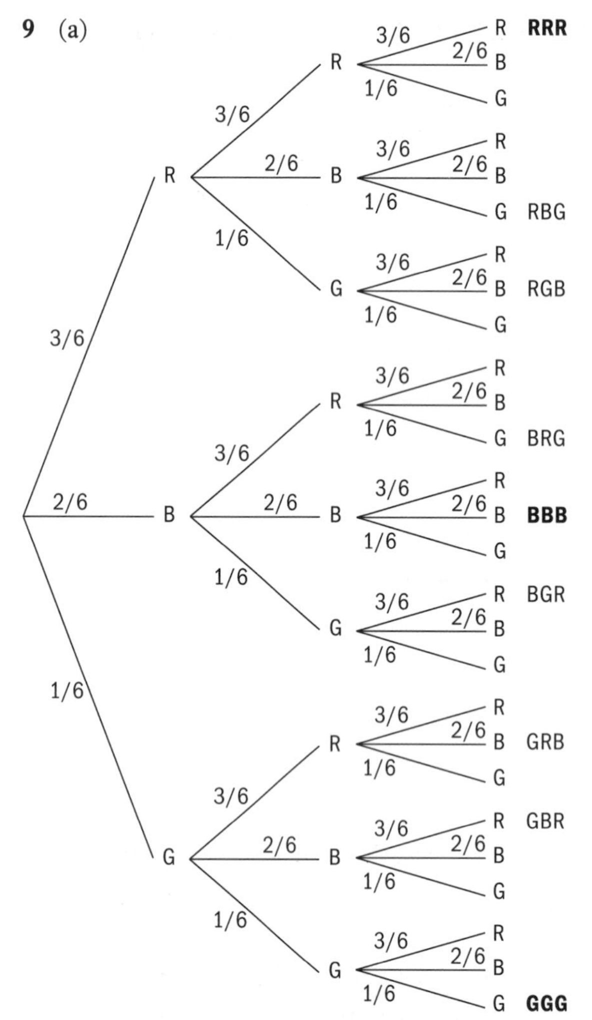

The outcomes listed on the right are the ones you need for the question.

(i) The probability of all three marbles being the same colour is

$(\frac{3}{6}\times\frac{3}{6}\times\frac{3}{6})+(\frac{2}{6}\times\frac{2}{6}\times\frac{2}{6})+(\frac{1}{6}\times\frac{1}{6}\times\frac{1}{6})$

$=\frac{27}{216}+\frac{8}{216}+\frac{1}{216}=\frac{36}{216}=\frac{1}{6}.$

(ii) The probability of all three marbles being different colours is

$6(\frac{1}{2}\times\frac{1}{3}\times\frac{1}{6})=6\times\frac{1}{36}=\frac{1}{6}.$

(b)

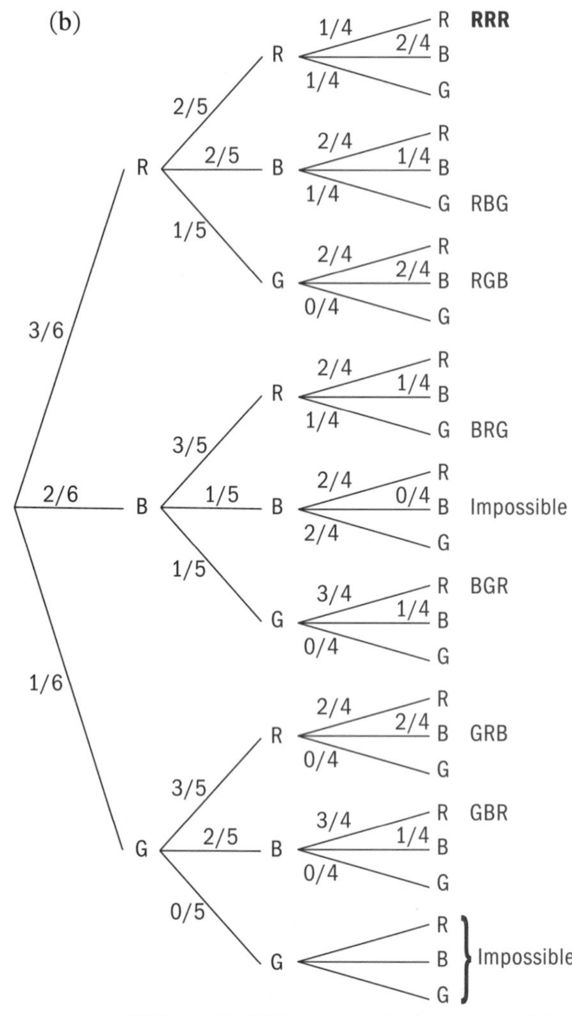

All the probabilities are marked above, but it is only necessary to work out the fractions along the branches you are interested in.

(i) The probability of all three marbles being the same colour is

$\frac{3}{6}\times\frac{2}{5}\times\frac{1}{4}=\frac{1}{20}.$

The only possible outcome is three reds.

(ii) The probability of all three marbles being different colours is

$(\frac{3}{6}\times\frac{2}{5}\times\frac{1}{4})+(\frac{3}{6}\times\frac{1}{5}\times\frac{2}{4})+(\frac{2}{6}\times\frac{3}{5}\times\frac{1}{4})$

$+(\frac{2}{6}\times\frac{1}{5}\times\frac{3}{4})+(\frac{1}{6}\times\frac{3}{5}\times\frac{2}{4})+(\frac{1}{6}\times\frac{2}{5}\times\frac{3}{4})$

$=6\times\frac{1}{20}$

$=\frac{3}{10}.$

10 (a) (i) $\frac{1}{2}$

(ii) $\frac{1}{6}$

If all three players are to move, the throw must be a '2'.

(iii) $\frac{1}{3}$

If nobody moves, the throw must be a '1' or '6'.

(b) $\frac{1}{6} \times \frac{1}{6} = \frac{1}{36}$

Ann needs a '4' and then a '6' to win in two throws.

(c) $\frac{2}{6} \times \frac{1}{6} = \frac{2}{36} = \frac{1}{18}$

Bob needs a '1' or '3' on his third throw and a '6' on his fourth throw.

11 (a) $\frac{1}{6} \times \frac{1}{6} = \frac{1}{36}$

The only way to score 12 in two throws is to throw two sixes, both with red dice.

(b) *'3' can be scored by throwing a '1' on the red dice followed by '2' on the blue, or '2' on the red dice followed by '1' also on the red. The probability is found by adding the individual probabilities together.*

$(\frac{1}{6} \times \frac{1}{4}) + (\frac{1}{6} \times \frac{1}{6}) = \frac{1}{24} + \frac{1}{36} = \frac{5}{72}$

More help or practice

Independent events ► Book Y4 pages 76 to 77
Tree diagrams ► Book Y4 pages 78 to 79
Adding probabilities ► Book Y4 pages 80 to 81
Dependent events ► Book Y4 pages 83 to 85
Conditional probability ► Book YX2 pages 8 to 12

Mixed handling data (page 90)

1 (a) (i) The cumulative frequencies are 130, 6473, 18 256, 23762, 25221, 25436, 25467.

(ii) The plotted points should be (2·5, 130), (3·5, 6473), (4·5, 18 256), and so on.

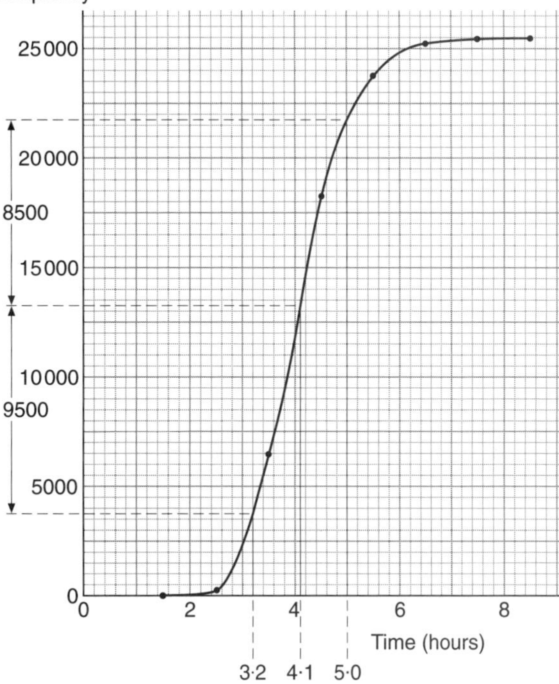

(b) Estimated mean $= \frac{104458}{25467} = 4{\cdot}1$ (to 1 d.p.)

Standard deviation $= 0{\cdot}9$ (to 1 d.p.)

(c) (i) Time interval is 4·1 to 5·0 hours.
The approximate number of competitors who finished within this range is 8500.
This gives a percentage of
$\frac{8500}{25467} \times 100\% = 33\%$.

(ii) Time interval is 3·2 to 4·1 hours.
The approximate number of competitors who finished within this range is 9500.
This gives a percentage of
$\frac{9500}{25467} \times 100\% = 37\%$.

Even on your own full-size graph, the vertical scale is very small which makes taking accurate readings difficult. As long as you showed your working, any answer in the range (i) 29–37% and (ii) 33–41% would gain full marks.

2 (a) 58% (b) 60% (c) 40%

(d) Treatments X and Y were equally successful. Patients who received either of the treatments had their chance of recovery improved by about 50%.

3 (a) 24 possible arrangements

If the first card is A, there are six possible arrangements of the other three cards:

A P Q X
A P X Q
A Q P X
A Q X P
A X P Q
A X Q P

Similarly for the other three cards.

(b) Divide by 2 because there are two Ps.

(c) 120 arrangements

A fifth card increases the number of arrangements by a factor of 5.

4 (a)

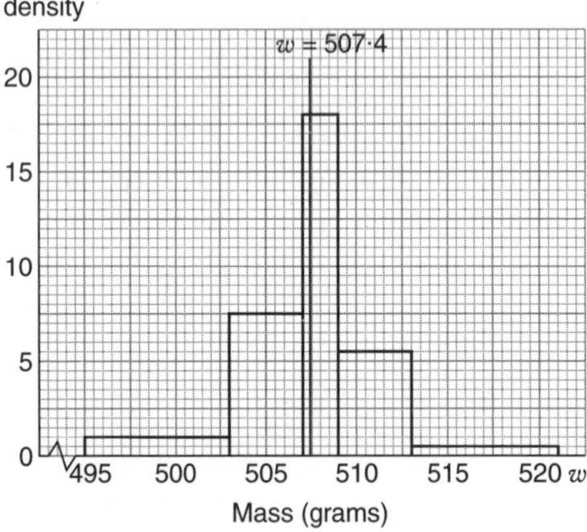

(b) 55%

As there are 100 packets, the percentage is simply given by the area to the right of the line $w = 507{\cdot}4$.

(c) The mid-interval values are 499 g, 505 g, 508 g, 511 g and 517 g. Using the usual notation (see page 80),

$$\Sigma fX^2 = 25\,746\,880$$

Standard deviation

$$= \sqrt{\Sigma \frac{fX^2}{100} - 507{\cdot}4^2} = \sqrt{14{\cdot}04} = 3{\cdot}7$$

(to 1 d.p.)

(d) $m - 2s = 507{\cdot}4 - 7{\cdot}4 = 500$

So an estimated 5 packets have a mass of less than C grams.

5 (a)

Price £p	Mid-interval value (fX)	Frequency (f)	fX
$1{\cdot}75 \le p < 2{\cdot}25$	2·0	1	2·0
$2{\cdot}25 \le p < 2{\cdot}75$	2·5	11	27·5
$2{\cdot}75 \le p < 3{\cdot}25$	3·0	24	72·0
$3{\cdot}25 \le p < 3{\cdot}75$	3·5	28	98·0
$3{\cdot}75 \le p < 4{\cdot}25$	4·0	14	56·0
$4{\cdot}25 \le p < 4{\cdot}75$	4·5	6	27·0
$4{\cdot}75 \le p < 5{\cdot}25$	5·0	3	15·0
$5{\cdot}25 \le p < 6{\cdot}25$	5·75	3	17·25
$6{\cdot}25 \le p < 7{\cdot}25$	6·75	4	27·0

Total 341·75

Estimated mean = £341·75 ÷ 94 = £3·64 (to 2 d.p.)

(b) The cumulative frequencies are 1, 12, 36, 64, 78, 84, 87, 90 and 94.

(c) Reading off the graph at 47 (or 47·5) gives a median price of approximately £3·40.

(d) *Either of these answers would be acceptable.*

The median is the best 'average' as it is not affected by very cheap or very expensive wines.

The mean is the best 'average' as it takes into account the price of all the wines.

(e) $\frac{24}{94} \times 100\% = 25{\cdot}5\%$ (to 1 d.p.)

Use the graph to find that about 24 bottles are priced at under £3.

MIXED AND ORALLY-GIVEN QUESTIONS

Mixed questions 1 (page 92)

1 *It often helps to start by marking the information you know on a sketch.*

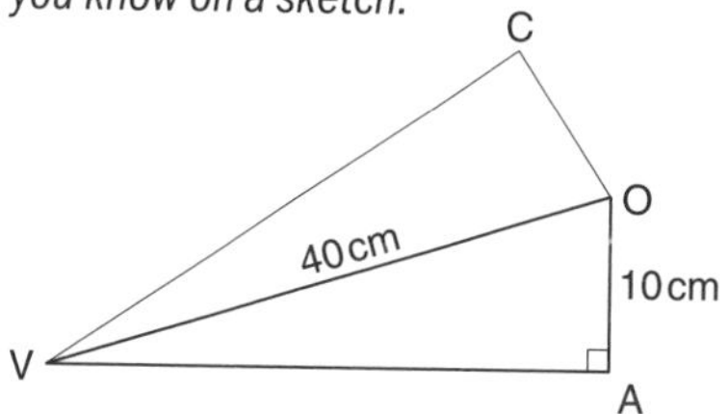

(a) (i) By Pythagoras, $VA^2 = 40^2 - 10^2 = 1500$
$VA = \sqrt{1500} = 38{\cdot}729\ldots$
$= 38{\cdot}7$ cm (to 1 d.p.)

(ii) $\cos AOV = \frac{10}{40} = 0{\cdot}25$
$\angle AOV = 75{\cdot}5°$ (to 1 d.p.)

(iii) The obtuse $\angle AOC = 2 \times 75{\cdot}5° = 151°$
The reflex $\angle AOC = 360° - 151° = 209°$
Area of sector OABC $= \frac{209}{360} \times \pi \times 10^2 \text{cm}^2$
$= 182 \text{cm}^2$ (to 3 s.f.)

(b) (i) PV and PO both bisect $\angle AVC$, so VPO must be a straight line.

(ii) $VP = VO - PO = 40 - (r + 10) = 30 - r$

(iii) $\angle VFP = \angle VAO = 90°$
$\angle PVF = \angle OVA$ (same angle)
So $\angle VPF = \angle VOA$ (angle sum of triangle)
The triangles have three equal angles and therefore must be similar.

(iv) Using the properties of similar triangles,
$\frac{r}{30 - r} = \frac{10}{40}$.
Multiplying to remove the fractions gives
$40r = 300 - 10r$
$50r = 300$
$r = 6$

2 (a) For all four pairs of values $w = \frac{36}{t}$ (or $wt = 36$)

(b) (i) 10 cm^2
You would gain full marks for an answer within ± 1 of this.

(ii) The volume of water flowing between $t = 1$ and $t = 4$, which is 50 m^3/hour.
The area of 1 square represents 5 m^3/hour.

(c) (i) Any answer between $^-8$ and $^-11$

(ii) The rate of flow is decreasing (by 9 m^3/h per hour).

3 (a) $x^2 - y^2 = (x - y)(x + y)$

(b) $9996 = 100^2 - 2^2 = (100 - 2)(100 + 2)$
$= 98 \times 102$
$= 2 \times 49 \times 2 \times 3 \times 17$
$= 2^2 \times 3 \times 7^2 \times 17$

4 (a) $AB^2 = AC^2 - BC^2$
$= (4^2 \times 6) - (6^2 \times 2)$
$= 96 - 72 = 24$
$AB = \sqrt{24} \text{cm} = 2\sqrt{6} \text{cm}$

(b) The ratio of the area of ΔAPQ to the area of ΔABC is 1 to 2. By the rules of enlargement, the ratio of AP to AB is 1 to $\sqrt{2}$. So
$\frac{AP}{AB} = \frac{1}{\sqrt{2}}$
$AP = \frac{AB}{\sqrt{2}} = \frac{2\sqrt{6}}{\sqrt{2}} = \sqrt{12} = 2\sqrt{3}$

5 (a) Both numbers are rational because they can be written as fractions.
(i) $5{\cdot}252525 = 5\frac{252\,525}{1\,000\,000}$ (ii) $5{\cdot}\dot{2}\dot{5} = 5\frac{25}{99}$

(b) (i) The circumference of a circle is given by $2\pi r$ so, since π is irrational and $2r = 4$ is rational, the circumference is also irrational.

(ii) The hypotenuse length is given by $\sqrt{(2^2 + 5)} = \sqrt{9} = 3$, so it is rational.

(iii) The perpendicular height is given by $\sqrt{(2^2 - 1)} = \sqrt{3}$, which is irrational.

6 (a) $16^{\frac{1}{2}} = 4, 4^{-1} = \frac{1}{4}, 32^0 = 1, (^-2)^4 = 16, 2^{-2} = \frac{1}{4}$
So $4^{-1} = \frac{1}{4} = 2^{-2}$

(b) $\frac{1}{6}$

(c) $4^x = 2^{2x} = 8 = 2^3$
$2x = 3$
$x = 1{\cdot}5$

Mixed questions 2 (page 94)

1 (a) In similar triangles, corresponding angles are equal.
So $\angle SQP = \angle SPR$.
Also $\angle SPQ + \angle SQP = 90°$.
So $\angle QPR = \angle SPQ + \angle SPR$ is a right-angle.

(b) (i) $b = ka$ and $kb = 3a$
(ii) Substituting $b = ka$ in the second equation gives $k^2a = 3a$ which simplifies to $k^2 = 3$.

(c) $RQ = a + kb = a + k^2a = 4a$
In triangle QPS,
$c^2 = a^2 + b^2$
$= a^2 + 3a^2$ *substituting* $b = \sqrt{3}a$
$= 4a^2$
so $c = 2a$.
But $RQ = 4a$, so $RQ = 2c$.

2 (a) (i) Two-fifths or 40% of £56250 = £22500
So $x + y = 22500$
(ii) $3x + 2y = 58500$
(iii) Solving the simultaneous equations gives $x = 13500$ and $y = 9000$.

(b) 0·5% of price = £271·25
100% of price = £271·25 × 200 = £54250

3 The arc length is given by $\frac{2x}{360} \times 2\pi \times 45 = 60$.
Solving this equation gives $x = 38{\cdot}2°$ (to 1 d.p.).
It is important to show all your working. In this question you would have gained a mark for writing down the circumference of the circle, even if you made a mistake in forming or solving the equation.

4 Maximum track length = 100·5 metres
Minimum time = 12·55 seconds
Greatest possible average speed
= 100·5 ÷ 12·55 m/s
= 8·0 m/s (to 1 d.p.)

5 (a) E (b) A (c) A (d) D

6 (a) $\pi rl = 2\pi rh$ *equating the curved surface areas*
$l = 2h$
$r^2 = l^2 - h^2$ *by Pythagoras*
$r^2 = 4h^2 - h^2$ *substituting* $l = 2h$
$= 3h^2$

(b) Volume of cone $= \frac{1}{3}\pi r^2h$
$= \frac{1}{3}\pi \times 3h^2 \times h = \pi h^3$
Volume of cylinder $= \pi r^2h = 3\pi h^3$

(c) (i) $3\pi h^3 - \pi h^3 = 16\pi$
$2\pi h^3 = 16\pi$
$h^3 = 8$
$h = 2$
(ii) $r^2 = 3h^2 = 3 \times 2^2 = 12$
$r = \sqrt{12} = 3{\cdot}46$ (to 3 s.f.)

(d) (i) $l = \sqrt{(r^2 + h^2)} = \sqrt{(12 + 4)} = 4$
It is important (and easier) to use the surd value of r.
(ii) $\pi rl = \pi \times \sqrt{12} \times 4$
Area = 43·5 cm² (to 3 s.f.)

(e) $\frac{\text{Area of shaded sector}}{\text{Area of whole circle}} = \frac{4\pi\sqrt{12}}{\pi \times 4^2} = \frac{\sqrt{12}}{4} = \frac{\sqrt{3}}{2}$

(f) $\frac{x}{360} = \frac{\sqrt{3}}{2}$
$x = 180\sqrt{3}$
$= 312°$ (to nearest degree)

Mixed questions 3 (page 96)

1 (a) 9°C

(b) (i) $^{-}10$°C (ii) 8 minutes

2 (a) Volume in $m^3 = \pi r^2 h$

$= \pi \times 3{\cdot}5^2 \times 10^{-8} \times 5{\cdot}8 \times 10^3$

$= 2{\cdot}23 \times 10^{-3}$ (to 3 s.f.)

(b) 0·00223 (to 3 s.f.)

3 (a) Total length

$= \frac{110}{360} \times 2\pi(3{\cdot}5 + 1{\cdot}5) + 2(3{\cdot}5 - 1{\cdot}5)$ m

$= 13{\cdot}6$ m (to 1 d.p.)

(b) Area $= \frac{110}{360} \times \pi(3{\cdot}5^2 - 1{\cdot}5^2)\,m^2$

$= 9{\cdot}599\ldots\,m^2 = 9{\cdot}60\,m^2$ (to 3 s.f.)

(c) Weight = area × 135 g = 1295·906… g

= 1300 g or 1·30 kg (to 3 s.f.)

Use the uncorrected value of the area when calculating the weight.

4 (a) $(31 \times 4{\cdot}5) + (30 \times 5{\cdot}8) = 313{\cdot}5$ hours

(b) $313{\cdot}5 \times 2{\cdot}44 = 764{\cdot}94$ litres

= 765 litres (to nearest litre)

(c) (i) £173 × £764·94 ÷ 900

= £147 (to nearest pound)

(ii) £147·038 … ÷ 61

= £2·41 (to nearest penny)

5 (a) Area of carpet in m^2

$= (6 - 2x)(4 - 2x) = \frac{1}{2} \times 24$

So $24 - 20x + 4x^2 = 12$

$4x^2 - 20x + 12 = 0$

$x^2 - 5x + 3 = 0$

(b) $x = 0{\cdot}697$ m or 69·7 cm (to 3 s.f.)

The second solution, $x = 4{\cdot}30$ m, does not make sense in this context.

6 (a) (i) 21·4 years (ii) 23·24 years

(iii) 19·8 years (iv) 2015

(b) (i) 164 to 165 cm (ii) 156·5 to 157·5 cm

(iii) 13 to 14 cm

(iv) The frequencies are 7, 20, 31, 17, 5

(v) $((7 \times 145) + (20 \times 155) + (31 \times 165) + (17 \times 175) + (5 \times 185)) \div 80$

= 164·125 cm

= 164 cm (to 3 s.f.)

Remember to show all your working.

Orally-given questions 1 (page 98)

1 5

2 23

3 40

4 20 cm

5 42 000

6 32%

7 594

8 $3n - 2$

9 0·55

10 34 m.p.h.

11 13 CDs

12 108°

13 £43·75

14 8

15 $\frac{9}{25}$

16 60 m^2 or 70 m^2

17 £9·00

18

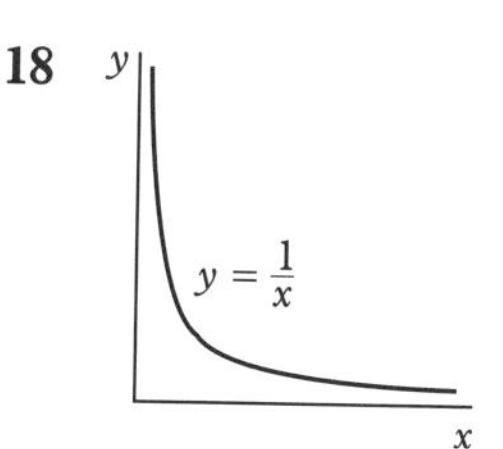

Orally-given questions 2 (page 99)

1 30

2 11

3 $^-19$°C

4 6·2

5 240 cm

6 5

7 6

8 43° each

9 $\begin{bmatrix} ^-4 \\ 3 \end{bmatrix}$

10 £19·37

11 $\frac{1}{3}$

12 n^2

13 315°

14

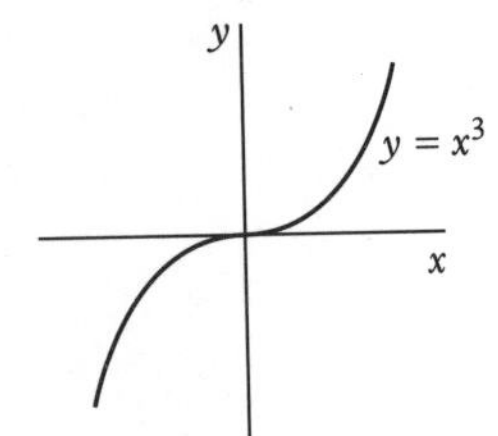

15 £85·50

16 500 km

17 $2x + 5 > x^3$ or $5 + 2x > x^3$ or $x^3 < 5 + 2x$

18 $\frac{580}{29^2} \approx \frac{600}{900} = \frac{2}{3}$ or 0·6 or 0·7